THE NATIONAL ACADEMIES

National Academy of Sciences
National Academy of Engineering
Institute of Medicine
National Research Council

The **National Academy of Sciences** is a private, nonprofit, self-perpetuating society of distinguished scholars engaged in scientific and engineering research, dedicated to the furtherance of science and technology and to their use for the general welfare. Upon the authority of the charter granted to it by the Congress in 1863, the Academy has a mandate that requires it to advise the federal government on scientific and technical matters. Dr. Bruce M. Alberts is president of the National Academy of Sciences.

The **National Academy of Engineering** was established in 1964, under the charter of the National Academy of Sciences, as a parallel organization of outstanding engineers. It is autonomous in its administration and in the selection of its members, sharing with the National Academy of Sciences the responsibility for advising the federal government. The National Academy of Engineering also sponsors engineering programs aimed at meeting national needs, encourages education and research, and recognizes the superior achievements of engineers. Dr. William A. Wulf is president of the National Academy of Engineering.

The **Institute of Medicine** was established in 1970 by the National Academy of Sciences to secure the services of eminent members of appropriate professions in the examination of policy matters pertaining to the health of the public. The Institute acts under the responsibility given to the National Academy of Sciences by its congressional charter to be an adviser to the federal government and, upon its own initiative, to identify issues of medical care, research, and education. Dr. Kenneth I. Shine is president of the Institute of Medicine.

The **National Research Council** was organized by the National Academy of Sciences in 1916 to associate the broad community of science and technology with the Academy's purposes of furthering knowledge and advising the federal government. Functioning in accordance with general policies determined by the Academy, the Council has become the principal operating agency of both the National Academy of Sciences and the National Academy of Engineering in providing services to the government, the public, and the scientific and engineering communities. The Council is administered jointly by both Academies and the Institute of Medicine. Dr. Bruce M. Alberts and Dr. William A. Wulf are chairman and vice chairman, respectively, of the National Research Council.

INITIAL REPORT

Panel on Formula Allocations

Thomas B. Jabine, Thomas A. Louis, and Allen L. Schirm, *Editors*

Committee on National Statistics

Division of Behavioral and Social Sciences and Education

National Research Council

NATIONAL ACADEMY PRESS
Washington, DC

NATIONAL ACADEMY PRESS   2101 Constitution Avenue, N.W.   Washington, DC 20418

NOTICE: The project that is the subject of this report was approved by the Governing Board of the National Research Council, whose members are drawn from the councils of the National Academy of Sciences, the National Academy of Engineering, and the Institute of Medicine. The members of the committee responsible for the report were chosen for their special competences and with regard for appropriate balance.

This study was supported by Contract/Grant No. RN 96131001 between the National Academy of Sciences and the U.S. Department of Education. Support of the work of the Committee on National Statistics is provided by a consortium of federal agencies through a grant from the National Science Foundation (Number SBR-9709489). Any opinions, findings, conclusions, or recommendations expressed in this publication are those of the author(s) and do not necessarily reflect the views of the organizations or agencies that provided support for the project.

International Standard Book Number 0-309-07580-7

Additional copies of this report are available from National Academy Press, 2101 Constitution Avenue, N.W., Lockbox 285, Washington, DC 20055. Call (800) 624-6242 or (202) 334-3313 (in the Washington metropolitan area). This report is also available online at http://www.nap.edu

Printed in the United States of America

Copyright 2001 by the National Academy of Sciences. All rights reserved.

Suggested citation: National Research Council (2001) *Choosing the Right Formula: Initial Report.* Panel on Formula Allocations, Thomas B. Jabine, Thomas A. Louis, and Allen L. Schirm, Editors. Committee on National Statistics. Washington, DC: National Academy Press.

PANEL ON FORMULA ALLOCATIONS

THOMAS A. LOUIS (*Chair*), RAND, Alexandria, Virginia
GORDON J. BRACKSTONE, Statistics Canada
LINDA GAGE, Demographic Research Unit, California Department of Finance, Sacramento
HERMANN HABERMANN, Statistics Division, United Nations, New York
ALLEN L. SCHIRM, Mathematica Policy Research, Inc., Washington, DC
BRUCE D. SPENCER, Department of Statistics, Northwestern University

VIRGINIA de WOLF, *Study Director*
DANELLE DESSAINT, *Project Assistant*
MARISA GERSTEIN, *Research Assistant*
THOMAS B. JABINE, *Consultant*

COMMITTEE ON NATIONAL STATISTICS
2001

JOHN E. ROLPH *(Chair)*, Marshall School of Business, University of Southern California
JOSEPH G. ALTONJI, Department of Economics, Northwestern University
LAWRENCE D. BROWN, Department of Statistics, University of Pennsylvania, Philadelphia
JULIE DAVANZO, RAND, Santa Monica, California
ROBERT M. GROVES, Joint Program in Survey Methodology, University of Maryland, College Park
HERMANN HABERMANN, Statistics Division, United Nations, New York
JOEL HOROWITZ, Department of Economics, The University of Iowa
WILLIAM KALSBEEK, Department of Biostatistics, University of North Carolina
RODERICK J.A. LITTLE, School of Public Health, University of Michigan
THOMAS A. LOUIS, RAND, Alexandria, Virginia
DARYL PREGIBON, AT&T Laboratories-Research, Florham Park, New Jersey
FRANCISCO J. SAMANIEGO, Division of Statistics, University of California, Davis
RICHARD L. SCHMALENSEE, Sloan School of Management, Massachusetts Institute of Technology
MATTHEW D. SHAPIRO, Department of Economics, University of Michigan, Ann Arbor

ANDREW A. WHITE, *Director*

Dedication

We note with sadness and respect the death of Wray Smith on May 19, 2000, less than one month after he played a major role in the April 2000 Workshop on Formulas for Allocating Program Funds that is the primary subject of this report. Wray had a distinguished career as a federal statistician. In 1976-1977, under the auspices of the Federal Committee on Statistical Methodology, he chaired the subcommittee that produced *Statistical Policy Working Paper 1: Report on Statistics for Allocation of Funds* (U.S. Office of Statistical Policy and Standards, 1978). That report broke new ground in two ways: it served as the prototype for the highly successful continuing series of *Statistical Policy Working Papers* and it identified the potential for improving the equity and effectiveness of formula allocation processes by paying greater attention to their statistical features. The recommendations in *Working Paper 1* have provided this panel with an important starting point for its deliberations.

Because of his groundbreaking contributions, Wray was asked to prepare and present an introductory background paper for the April 2000 Workshop (Smith and Parker, 2000), an assignment that he fulfilled with his usual dedication and skill. We dedicate this report to his memory.

Acknowledgments

Many people contributed their time and expertise to the development of this report of the Panel on Formula Allocations. Financial support was provided by the U.S. Department of Education and Committee on National Statistics. This support made possible the panel and staff work that serves as the basis for this report. Daniel Kasprzyk of the National Center for Education Statistics who served as project officer for the study for the U.S. Department of Education, was most helpful in facilitating the panel's work throughout the project.

Contributions of the presenters and discussants at the workshop were informative and invaluable. Many other individuals offered important comments and suggestions at our subsequent meetings as well: Jim Adams, U.S. General Services Administration; Chip Alexander, U.S. Census Bureau; Susan Binder, U.S. Department of Transportation, Federal Highway Administration; Sandy Brown, U.S. Department of Education; Pasquale DeVito, Board on Testing and Assessment; Thomas Downes, Tufts University; Jerry Fastrup, U.S. General Accounting Office; Gregory Frane, U.S. Department of Education; Jerry Keffer, U.S. Census Bureau; Kaeli Knowles, Board on Testing and Assessment; Cindy Long, U.S. Department of Agriculture, Food And Nutrition Service; David McMillen, U.S. House Government Reform and Oversight Committee; Wayne Riddle, Congressional Research Service; Marjorie Siegel, U.S. Department of Housing and Urban Development; William Sonnenberg, National Center for Education Statistics; Max Storrs, U.S. Department of Health and Human Services; Cynthia Taeuber, U.S. Census Bureau; Karen Wheeless, U.S. Census Bureau; and Albert Woodward, Substance Abuse and Mental Health Services Administration.

The panel is deeply indebted to Thomas Jabine, whose thoughtful guidance and intellectual contributions permeate this report. I give special thanks to the panel members, who have donated their time and expertise. Staff of the Committee on National Statistics provided key editorial, organizational, and research contributions. Heather Koball served as study director in preparing for the workshop. After the workshop, Andy White developed institutional and financial support for the panel and provided a bridge to Virginia de Wolf, our current study director. She and Danelle Dessaint assisted in organizing and preparing this report, and Marisa Gerstein carried out important research tasks.

This report has been reviewed in draft form by individuals chosen for their diverse perspectives and technical expertise, in accordance with procedures approved by the Report Review Committee of the National Research Council. The purpose of this independent review is to provide candid and critical comments that will assist the institution in making the published report as sound as possible and to ensure that the report meets institutional standards for objectivity, evidence, and responsiveness to the study charge. The review comments and draft manuscript remain confidential to protect the integrity of the deliberative process.

We thank the following individuals for their participation in the review of this report: John Czajka, Mathematica Policy Research, Inc., Washington, DC; Thomas Gabe, Domestic Social Policy Division, Congressional Research Service, U.S. Congress; Joel Horowitz, Department of Economics, University of Iowa; and James Wyckoff, Graduate School of Public Affairs, State University of New York at Albany.

Although the reviewers listed above have provided many constructive comments and suggestions, they were not asked to endorse the conclusions or recommendations nor did they see the final draft of the report before its release. The review of this report was overseen by Joel Greenhouse, Department of Statistics, Carnegie Mellon University. Appointed by the National Research Council, he was responsible for making certain that an independent examination of this report was carried out in accordance with institutional procedures and that all review comments were carefully considered. Responsibility for the final content of this report rests entirely with the authoring panel and the institution.

Thomas A. Louis, *Chair*
Panel on Formula Allocations

Contents

Foreword	xiii
PART I WORKSHOP REPORT	1
1 Formula Allocation Processes: An Overview	3
2 Case Studies	12
3 Effects on Formula Outputs of Errors in Formula Inputs	37
4 Roundtable and Concluding Sessions	45
PART II PANEL REPORT	53
5 Themes and Issues	55
6 Anticipated Panel Activities	61
References and Bibliography	70
APPENDIX A Workshop Agenda and Participants	75
APPENDIX B Biographical Sketches of Panel Members and Staff	83

Foreword

Each year, formulas are used to allocate well over $200 billion of federal funds to state and local governments via more than 160 federal programs designed to meet a wide spectrum of economic and social objectives. These programs address societal goals, such as improving educational outcomes and accessibility of medical care, and are designed to equalize fiscal capacity to address identified needs. An early example of such a formula was the Morrill Act of 1862, which allotted to each state 30,000 acres of public land for each of its senators and representatives in Congress. The land was to be sold and the proceeds used to establish one or more institutions of higher learning.

Such formulas are developed in the context of a complex political process. Use of a formula, as opposed to arbitrary specification of the amount to be given to each recipient jurisdiction, facilitates informed debate about the allocation process by providing documentation of assumptions and computations. Furthermore, a formula offers legislators an effective way of explaining the allocation process to their constituents. However, as discussed in this report, when funds are allocated according to a formula, there is no guarantee that objectives will be fully met. In particular, properties of data sources and statistical procedures used to produce formula inputs can interact in complex ways with formula features to produce consequences that may not have been anticipated or intended.

There is a long history of attention to these matters, but many of the issues identified and recommendations made are still of central importance and have been only partially addressed. To reenergize attention to these persistent issues, the Committee on National Statistics (CNSTAT) convened a two-day Workshop on Formulas for Allocating Program Funds on April 26-27, 2000. The workshop focused on statistical issues that arise in the development and use of formulas for allocating federal funds to state and local governments in programs with a wide spectrum of objectives. Its agenda included an overview of formula allocation programs and relevant data sources, case studies, presentations on methodology, and a roundtable discussion. Presenters and other workshop participants included formula allocation program managers, economists, statisticians, and demographers from federal and state government agencies, universities, and independent research organizations. The goals of the workshop planners were to issue a report that would make a significant contribution to the field and to lay the groundwork for a subsequent panel study.

The workshop was a direct outgrowth of a previous study by the CNSTAT Panel on Estimates of Poverty for Small Geographic Areas. That panel, established under a 1994 act of Congress, began its work with a very specific mission: to evaluate the suitability of the U.S. Census Bureau's small-area estimates of poor school-age children for use in the allocation of funds to counties and school districts under Title I of the Elementary and Secondary Education Act. In carrying out their assignment, panel members came to realize that the properties of data sources and statistical procedures used to produce formula estimates, interacting with formula features such as thresholds and hold-harmless provisions, can produce consequences that may not have been anticipated or intended (See Chapter 3 for specific examples). It also became evident that there is a trade-off between the goals of providing a reasonable amount of stability in funding from one year to the next and redirecting funds to different jurisdictions as true needs change. In one instance, for example, the annual appropriation included a 100 percent hold-harmless provision, ensuring that no recipient would receive less than the year before (For details, see "Title I of the Elementary and Secondary Education Act" in Chapter 2). However, there was no increase in the total appropriation, with the result that new estimates showing changes in the distribution of program needs across areas had no effect on the allocations.

Situations like this can arise not only in the Title I education allocations, but also in the many other formula allocation programs under which large amounts of federal and state funds are distributed to local governments for defined purposes. In considering the panel's conclusions,

FOREWORD

CNSTAT decided that it was important to explore these issues in a broader context, starting with the workshop and proceeding to a more comprehensive panel study.

Following the April 2000 workshop, the Panel on Formula Allocations was formed in fall 2000. The panel's tasks are to refine and follow up on the important issues identified in the workshop, conduct case studies and methodological investigations, obtain input from individuals who design and implement programs using formula allocation, and to develop findings, recommendations, and guidelines relating to these issues. To these ends, a planning meeting was held on December 11-12, 2000; the panel then met on January 11-12, February 15, and April 19-20, 2001, to develop a work plan, review information, and obtain input from individuals involved in developing and administering funds allocation programs. The panel will hold additional meetings during the remainder of 2001 and in 2002.

At its first meeting in January 2001, the panel decided to include a summary of the April 2000 workshop in this initial report; the issues identified by workshop participants will be high on the agenda for the panel's work. Part I of this report is the workshop summary. Part II synthesizes the principal themes from the workshop and the panel's initial meetings, highlights the principal issues the panel intends to address, and outlines anticipated panel activities.

We invite and encourage feedback on this report and the panel's proposed activities. Please address comments to Virginia de Wolf, study director, at the Committee on National Statistics (2101 Constitution Avenue, N.W., Washington, DC 20418, phone 202-334-3023; fax 202-334-3751; email vdewolf@nas.edu).

> John E. Rolph, *Chair*
> Committee on National Statistics

PART I

Workshop Report

This part of the report provides an account of the presentations and discussions at the workshop (see the agenda in Appendix A). The first three chapters cover the overview, case studies, and methodological sessions, respectively. Chapter 4 summarizes the issues discussed in the roundtable and concluding sessions, with emphasis on the identification of questions that might be addressed in a panel study.

1

Formula Allocation Processes: An Overview

The use of formulas to allocate federal and state funds to subordinate jurisdictions is part of a broader process of government-to-government transfer of funds. Uses of such funds by the recipients may be unrestricted, as in the General Revenue Sharing Program of the 1970s and 1980s, or they may be limited to specific purposes, for example to provide medical care and improve the education of children in poor families, assist persons to end their dependence on welfare, revitalize economically depressed cities and neighborhoods, or provide assistance to localities disproportionately affected by the HIV epidemic. At the federal level, the U.S. Congress determines how much money will be distributed and for what purposes. For some programs, the Congress appropriates a fixed total amount each year to be allocated among states or other recipients; for others, such as Medicaid, amounts may be specified as a certain proportion of all qualified expenditures by a state or other jurisdiction. In the former case, a formula dictates how much of the total goes to each recipient; in the latter case, a formula determines what proportion of each jurisdiction's amount will be matched by the federal government.

Wray Smith, of the Harris Smith Institutes, opened the workshop by presenting "An Overview of Formulas for Allocation of Funds" (Smith and Parker, 2000). He noted that nearly $200 billion of federal funds are distributed annually to states and other units of government under formula allocation programs. Amounts have more than doubled in real terms over the past 25 years; a 1975 study estimated that $35.6 billion was allocated under grants using population or per capita income as formula components.[1] In fiscal year 1998, Medicaid was by far the largest formula allocation program, with $101.2 billion disbursed. Highway planning and construction grants came next, with $19.8 billion, followed by allocations under Title I of the Elementary and Secondary Education Act, with $7.8 billion. The U.S. General Services Administration's 1998 *Formula Report to the Congress* lists a total of 340 programs; however, some of these do not have formula provisions but have optional or required matching or cost-sharing provisions.

Formula allocation programs are characterized by the allocation of money to states or their subdivisions in accordance with a distribution formula prescribed by law or administrative regulation, for activities of a continuing nature not confined to a specific project. For some programs, the distribution formula used is a closed mathematical expression; for others, iterative processes are used to arrive at the final allocations. *Block grant programs* are a subset of formula allocation programs in which the recipient jurisdiction has broad discretion for the application of funds received in support of such programs as community development or the prevention and treatment of substance abuse, which are specified in the enabling legislation. *Matching grant programs*, such as Medicaid and certain transportation programs, require that the recipient state provide a matching percentage of funds from state sources.

ELEMENTS INCLUDED IN ALLOCATION FORMULAS

Elements included in formulas vary widely among the programs currently active. Most programs use one or more of the following:

- A direct or indirect measure of *need*, such as the number of school-age children in poverty, the number of overcrowded housing units in an area, or the number of reported cases of AIDS.
- A measure of the *capacity* or *capability* of an area to meet the need

---

[1] The 1975 study, "Use of Data on Population in Federal Grants-in-Aid to State and Local Government in Fiscal 1975" was prepared by Charles Ellett of the Statistical Policy Division, U.S. Office of Management and Budget, and is cited in U.S. Office of Statistical Policy and Standards (1978). The 1975 and current figures may not be precisely comparable in terms of the types of formula allocation programs included.

from state, local, or private funds. Typical measures used are per capita income and total taxable resources.

• A measure of *effort*, that is, the amount of available local resources actually devoted to meeting the need. In the Medicaid formula, for example, this would be a state's total eligible medical expenses under its Medicaid program. In the Title I education program it is the state's average per pupil expenditure (bounded by 80 and 120 percent of the national average).

• An index of *costs* incurred in meeting program needs in an area, such as an index of wages paid to workers in the health care industry.

In addition to these and other formula elements, the allocation rules may include one or more of the following features:

• A *threshold*, which calls for some minimum level of need before an area is eligible for any funds at all under the program. In some programs, thresholds are used to target resources to the areas with the greatest need.

• A *minimum amount* to be received by each state or other jurisdiction.

• A *hold-harmless* provision, which limits decreases in amounts received by areas from one time period (usually a fiscal year) to the next.

The inclusion of such special features sometimes requires use of relatively complicated iterative procedures to determine the allocation of a fixed total appropriation to eligible jurisdictions.

DATA SOURCES

Specific data sources for formula elements may or may not be identified in enabling legislation for formula grant programs. Population (total or for defined age groups) is an element in many formulas and may come from the most recent decennial census or from the U.S. Census Bureau's current population estimates. Income data may come from the decennial census, the Current Population Survey (CPS), or personal income statistics compiled by the Bureau of Economic Analysis. Possible future sources of income and poverty data are the U.S. Census Bureau's Survey of Income and Program Participation and its American Community Survey (ACS), which is currently in the developmental stage (see discussion in Chapter 4).

Data from administrative sources may also be used as inputs. As described in Chapter 2, the county estimates of poor school-age children used in the Title I education allocations are model-based estimates that supplement decennial census and CPS data with inputs based on individual income tax returns and records of participation in the Food Stamp Program.

Several considerations influence choices among alternative sources of input data for elements included in a formula:

- The *conceptual fit* between currently available data and the formula elements, as defined in enabling legislation or administrative regulations. If the definitions of the elements or program goals lack specificity, evaluation of the fit may require subjective judgments.
- The *level of geographic* detail for which data are provided. The decennial census can provide estimates for areas as small as school districts (although with substantial sampling variability for the smaller districts), whereas estimates from the Survey of Income and Program Participation are limited to census divisions and a few large states.
- The *timeliness* of the data, that is., the elapsed time between the reference period for the estimates and the period for which the allocations are being made. Here the decennial census data are at an obvious disadvantage compared with continuing or periodic sample surveys and administrative record sources.
- The *quality* of the data, as measured in terms of sampling variability and bias.
- The *cost* of collecting or compiling new data to provide inputs to the formula. Benefits from improvements in conceptual fit or other aspects of data quality have to be weighed against cost. Even when existing data sources are used, there may be significant costs of obtaining data in a format suitable for the allocation process.

Clearly, there are many trade-offs among these considerations, and it is likely that no one data source will be superior to the others on all counts. One solution to this dilemma may be the use of model-based estimates that combine inputs from several different sources.

Martin David of the University of Wisconsin, a discussant in the opening session, elaborated on these trade-offs, using the Title I education allocations as an illustration. Comparing alternative sources of income data, he noted that the most comprehensive data on income by source come from the Survey of Income and Program Participation, but that it has the smallest sample size. The CPS data are somewhat less detailed but are based on a larger sample. Individual income tax data are not subject to sampling error but cover only about 90 percent of the total population. Their utility could be improved if they were coded to the county and school district levels. Decennial census data cover a larger proportion of the population and provide more geographic detail, but they lack timeliness and are subject to greater underreporting of some types of income. Regarding conceptual fit, he argued that the current official definition of poverty could be improved by adopting proposed revisions that include in-kind income as a resource and exclude taxes paid (see National Research Council, 1995).

HOW FORMULAS ARE DEVELOPED AND ADMINISTERED

David McMillen, a staff member of the U.S. House Committee on Government Reform and Oversight, discussed practical considerations that affect the development of legislation to allocate federal funds to states and localities. He stated that inclusion of a funding formula in a bill makes the legislative process more difficult from beginning to end. At each stage, starting with the committee that first considers the bill and proceeding through votes in both houses and deliberations in conference committees, the sponsors and drafters of the legislation must take into account how the members most influential at each stage will fare in allocations based on the proposed formula. As with any type of legislation, there must be a majority of members voting for it as it moves to final passage. However, funding formulas place in stark contrast those who are advantaged and disadvantaged by the legislation. In simplest terms, how does one persuade a member of Congress to vote for legislation that will disadvantage his or her constituents? There must be more winners than losers, and key members of the power structure must not perceive that the outcome will be unfavorable to them.

In the Senate, where an individual senator can block legislation from coming to the floor, it is important that members representing small states not feel that the allocation formula treats them unfairly. All members of Congress must face the reality that they will be held accountable for every vote they cast and that their constituents will be more inclined to judge them on how their state or district fared in the allocation than on the overall goals of the program. Thus, formulas that finally emerge from the political process may represent compromises between substantive program goals and the need to generate the required number of votes at each stage.

In some extreme cases, it has proven to be too difficult to devise a formula that would give the desired percentages to each state, so the actual percentages have been specified in the legislation.

In response to a question, McMillen said that members of Congress and their staffs frequently request that the Congressional Research Service and the U.S. General Accounting Office provide information used to evaluate proposed formulas. However, Congress has relatively little contact with the statistical community in this context. One exception was the initiative by Congressman Tom Sawyer of Ohio which led to the replacement of outdated census data by more current small-area estimates of income and poverty in the Title I education allocations (see Chapter 2 for more details).

Program agencies in the Executive Branch also play a significant role in the funding allocation process and in the formulation of rules and regulations that govern how the funds are used by the jurisdictions that receive them. In a few instances the agencies, following general guidelines in legislation, develop the specifics of the allocation formula or process. Funds are sometimes provided for agencies to conduct or sponsor research to determine to what extent the allocations have led to the achievement of program goals and to develop recommendations for improved formulas and better data sources.

At the state and local levels, authorities are generally required to follow prescribed administrative procedures in order to receive the funds that have been allocated to them and to account for their use. Typically, legislation or regulations allow for some proportion of the total funds to be used for such administrative purposes.

PREVIOUS STUDIES OF THE STATISTICAL ASPECTS OF ALLOCATION FORMULAS

The paper by Smith and Parker that was presented at the workshop summarized selected previous studies that they considered relevant to its themes. The first four studies related to the General Revenue Sharing (GRS) Program. Between 1972 and 1986, the GRS program allocated federal funds to approximately 39,000 local jurisdictions using a formula based on population and per capita income:

• The *General Revenue Sharing Data Study* by the Stanford Research Institute (1974). This pioneering study focused on questions of the degree to which equitable allocations to states, and thence to localities, were dependent on the quality of the data used in the GRS allocation formulas. Several methods of improving the timeliness and accuracy of the estimates were recommended. The study addressed difficulties in making the estimates for small jurisdictions that were mandated by the GRS legislation.

- A 1975 study of alternative formulas for the GRS program undertaken by the Center for Urban and Regional Study at Virginia Polytechnic Institute and State University under a grant from the National Science Foundation. The study recommended inclusion of a poverty factor in the intrastate allocation formula and allocations on a per capita basis for units of government for which reliable estimates of income and poverty were unavailable.

- The 1980 report of the CNSTAT panel on Small-Area Estimates of Population and Income (National Research Council, 1980). This panel was asked to evaluate the U.S. Census Bureau's procedures for making postcensal estimates of population and income, a task largely motivated by the use of these estimates in the GRS program. The panel made several recommendations for improving the estimates. It also recommended that some limits be imposed on congressional and other requirements for the U.S. Census Bureau to make such estimates for very small areas.

- A staff report of the Federal Reserve Bank of Minneapolis (Stutzer, 1981). The GRS allocation formula included an element designed to reward localities with higher tax efforts (a ratio of state and local tax collections to total personal income). This study used simulation methods to evaluate the effects of such provisions on the recipient government's tax effort, spending levels, and welfare.

Two major studies of the statistical aspects of allocation formulas appeared toward the end of the 1970s. The first was *Statistical Policy Working Paper 1, Report on Statistics for Allocation of Funds*, prepared by a subcommittee of the Federal Committee on Statistical Methodology (chaired by Wray Smith). The subcommittee's goal was "to study, from the statistical standpoint, possible principles or guidelines which could be used to insure that the intent of Congress is fulfilled in the allocation of federal funds" (U.S. Office of Statistical Policy and Standards, 1978:1). The report included 5 case studies selected from the 10 largest programs using population and per capita income data as formula components. It considered the problem of measuring population, capability, and effort and took into account the effect of constraints and hold-harmless provisions on formula performance. It also discussed the special problems that arise in administering allocations to small areas. The subcommittee recommended that the goals of each allocation program be specified as clearly as possible and made several recommendations designed to assist program designers and drafters of legislation in meeting these goals more effectively.

The second broad study of these issues was conducted by the Center for Governmental Research (1980) under a grant from the National Science Foundation. This project was primarily concerned with "developing and applying analytical tools to evaluate the distributional and equalization effects of federal grant-in-aid formulas and to improve formula performance." The project final report called for several modifications of existing formulas, including "elimination of dual formula systems, updating of data elements, elimination of constraints, adjustments for cost-of-government differentials, and use of income per need unit ratios."

An article by Spencer (1982a) examined interactions between statistical issues and the technical and political aspects of formula design. Spencer recommended further statistical and policy research to explore a series of trade-offs: the use of simple versus more sophisticated formula components, the use of general-purpose data versus data produced primarily for use in allocation programs, the use of abrupt versus gradual eligibility thresholds, and updating versus not updating the statistical variables used in formulas. He recommended that statisticians and policy analysts collaborate in research on these issues.

The Smith and Parker paper also reviewed three studies conducted in the 1990s. A study by the Urban Institute reviewed a broad range of options for *Reforming the Medicaid Matching Formula*. It discussed several elements that could be included in the formula, "particularly emphasizing a broader-based measure of fiscal capacity, adding cost of living and cost of health care adjustments, and adding a new component that would incorporate health care needs" (Blumberg et al., 1993:vi).

In 1996 the U.S. General Accounting Office, at the request of the Committee on the Budget, U.S. House of Representatives, conducted a study that focused on "the extent to which the grant system succeeds in two objectives frequently cited by public finance experts: (1) encouraging states to use federal dollars to supplement rather than replace their own spending on nationally important activities and (2) targeting grant funding to states with relatively greater programmatic needs and fewer fiscal resources" (U.S. General Accounting Office, 1996:1). The report, which was called *Federal Grants Design Improvements Could Help Federal Resources Go Further*, concluded, "for the most part, the federal grant system does not encourage states to use federal dollars as a supplement rather than a replacement for their own spending on nationally important activities" (p.2) but added that not every grant is intended to do so. The report also concluded, "federal aid is not targeted to offset these fiscal imbalances. Consequently, lower income states face greater fiscal strain in financing federally aided services than higher income states with lower measurable needs" (p.2).

Finally, Smith and Parker described a report from the RAND Drug Policy Research Center, *Review and Evaluation of the Substance Abuse and Mental Health Services Block Grant Allocation Formula* (Burnham et al., 1997). These grants were the subject of the third case study presented and discussed at the workshop (see Chapter 2).

2

Case Studies

Four federal formula grant programs were selected to provide concrete illustrations of statistical issues that arise in the development and application of funding formulas. The Medicaid program, although its basic formula differs substantially from those of most other programs, was chosen for the first case study because it currently accounts for more than half of all federal funds disbursed annually via formula grants. The second case study relates to grants under Title I of the Elementary and Secondary Education Act. The third case study relates to block grants administered by the Substance Abuse and Mental Health Services Administration (SAMHSA), and the fourth covers allocation of funds in the Special Supplemental Nutrition Program for Woman, Infants, and Children (WIC). The allocation formulas used in all of these programs have been modified over time, partly on the basis of evaluation studies and research, in attempts to improve the equity of the allocations and their effectiveness in meeting program goals.

MEDICAID

In the opening session of the workshop, Jerry Fastrup, of the U.S. General Accounting Office, described the Medicaid program and the formula used to provide federal funds to match expenditures by the states under the program. Medicaid is a jointly funded federal-state program providing health care to low-income persons under eligibility standards set by the states, subject to federal regulations. State expenditures are partially reimbursed by the federal government. The proportion reimbursed to each state is determined according to a formula for the federal medical assistance percentage (FMAP):

$$FMAP = 1.00 - 0.45 \times \left(\frac{StatePCI}{U.S.PCI}\right)^2$$

where *PCI* = per capita income. FMAP is subject to a minimum of 50 percent and a maximum of 83 percent. The values of per capita income are based on a 3-year average of personal income estimates compiled by the Bureau of Economic Analysis. The lower a state's per capita personal income in relation to the national value, the higher is the proportion of its Medicaid costs that will be reimbursed by the federal government. Squaring the ratio magnifies this effect. The highest value of the FMAP in fiscal year 2000 was Mississippi's 76.8 percent. Ten states received the minimum 50 percent reimbursement.

Fastrup explained that the per capita income ratio in the formula serves a dual purpose. It is indirectly related to the relative number of low-income persons in the state, which is an indicator of the target population that the program is intended to serve (need), and also to the state tax base, that is, its ability to raise revenues to meet program needs (capacity). He said that there may be better indicators that could be used in the formula. An estimate of the number of persons in poverty might be a more direct indicator of program need. For capacity, estimates by state of total taxable resources that are prepared by the U.S. Department of the Treasury and used in the SAMHSA block grant formula would be a more direct measure. Further improvements could be achieved by introducing measures related to cost differentials by state. Such measures would take account of differentials in the costs of providing medical care and also in the proportion of the target population who are elderly and therefore require more resources per person.

Fastrup presented results of a simulation using a revised formula that included such elements. Based on this formula, the states of California and New York, which both have per capita incomes above the national average, were among those with the lowest capacity in relation to need. He then presented an analysis of the equalizing effects on relative capacity, by state, of federal aid based on the current FMAP formula. This analysis showed that more than half of the states that were above the national average in terms of their capacity to fund Medicaid without federal aid were farther above it with federal aid (disequalization) and that several that had been below average were taken well above the average (overequalization). A handful of states, including New York, California, and the District of Columbia, were farther *below* the national average when federal aid under the Medicaid program was taken into account. He concluded his presentation by pointing out that a large majority of states would lose funding if a more appropriate formula were used; hence, the likelihood of such a formula being adopted is small.

TITLE I OF THE ELEMENTARY AND SECONDARY EDUCATION ACT

Sandy Brown, of the U.S. Department of Education, described the history and scope of the program, the processes used in allocating funds, and the changes that have been made during the past decade as a result of legislation designed to bring about use of more timely data in the allocation process.

The Title I grants to school districts were first authorized in 1965 as a part of the Johnson administration's Great Society program. Their objective is to provide financial assistance to school districts and schools with high numbers or percentages of poor children to help meet the educational needs of children who are failing, or most at risk of failing, to meet challenging academic standards. The Title I program is the largest elementary and secondary education program of the federal government, with a current budget of almost $7.9 billion. Based on 1996-1997 data, Title I accounted for about 2.5 percent of total U.S. revenues from all sources that were devoted to elementary and secondary education. About 92 percent of all school districts receive Title I funds, and about 27 percent of all children ages 5 to 17 are affected by Title I services. Currently, there are two kinds of grants being made: basic grants, which account for about 84 percent of the total, and concentration grants, which are designed to supplement funding for school districts with heavy concentrations of poor children.

For each fiscal year, the grants to individual school districts[1] are deter-

---

[1]Prior to the 1999-2000 school year, allocations to counties were determined by the Department of Education and the allocations to school districts were determined by state education agencies. Starting with the 1999-2000 school year, the allocations to school districts are determined directly by the Department of Education, except that states have the option to group school districts with a population of less than 20,000 and determine allocations to these districts, using a procedure approved by the Department of Education.

mined by allocation formulas that are applied to the total amounts appropriated by Congress for basic and concentration grants. The allocation formulas for both kinds of grants are based primarily on estimates of the number of children ages 5 to 17 living in families below the official poverty line. They also use state per pupil expenditures as a proxy measure of the cost of educational services. Both kinds of grants are subject to eligibility thresholds. To qualify for a basic grant, a school district must have at least 10 "formula" children[2] and the number of formula children must exceed 2 percent of the school age population. To qualify for a concentration grant, either a district must have more than 6,500 formula children, or the number must exceed 15 percent of the total school-age population. There are also state minimums whose effect is to give a few small states somewhat larger amounts than they would otherwise receive. Title I also includes a hold-harmless provision for basic grants, under which each district is guaranteed to receive a specified minimum percentage of what it received the year before. These guaranteed percentages vary according to a step function based on the percentage of eligible children in the school age population.

The application of the formulas for the two kinds of grants is an iterative process. First, estimates of the number of eligible children are multiplied by 40 percent of state per pupil expenditure to produce an "entitlement" figure for each school district. These figures are then prorated to add to the total appropriation for that type of grant. Then the state minimums and, for basic grants, the hold-harmless provisions are applied, with further prorating of amounts remaining after these provisions are satisfied. Special features that affect the final allocations include:

• A provision of the statute that allows a state to reallocate the total amount that the Department of Education has allocated to its school districts with fewer than 20,000 people. A state that exercises this option can use alternate data sources, such as school lunch program data, for this purpose.

• Changes in school district boundaries. The universe of school districts on which the Department of Education bases its allocations lags the currently existing universe by about five years, so the states in which there

---

[2]"Formula" (eligible) children include children ages 5 to 17 in families with income below the poverty level, children in foster homes, children in families above the poverty level that receive benefits under the Temporary Assistance to Needy Families program, and children in local institutions for neglected and delinquent children.

have been changes must adjust the department's allocation by districts to account for boundary changes and the creation of new school districts.

• The creation of charter schools. In some states, individual charter schools are treated like school districts but draw students from one or more existing school districts.

Sandy Brown's presentation covered the procedures for the allocation of funds to school districts; he did not cover the procedures used by school districts to allocate the funds they receive to individual schools.

He explained the steps that were taken during the 1990s to bring about the use of more up-to-date estimates of the number of school children in poverty. The reauthorization of the Elementary and Secondary Education Act in 1994 required that the Department of Education use, for its 1997-1998 school year allocations, the U.S. Census Bureau's updated estimates of poor school-age children by county, unless the secretaries of education and commerce were to determine that these estimates were "inappropriate or unreliable" based on a National Research Council review that was authorized by the legislation. The Panel on Estimates of Poverty for Small Geographic Areas reviewed the U.S. Census Bureau's county estimates for 1993 and recommended that they be used in the allocation process. The estimates were used in the allocation to counties,[3] but at that stage it was left to the states to determine the suballocation of funds to school districts. Subsequent reports by the panel supported the use of revised 1993 county estimates for the 1998-1999 school year allocations (National Research Council, 1998) and the use of the U.S. Census Bureau's 1995 estimates by school district for the allocation of 1999-2000 Title I funds directly to school districts (National Research Council, 1999a).

Brown then discussed how changes during this period affected the amounts received by states and school districts. Factors that affect the allocations are:

• Changes in the estimated number of eligible children in each state, county, and school district. The estimated national poverty rate declined from 21.2 percent in 1993 to 19.5 percent in 1995, but there were in-

---

[3]For the fiscal year 1997 allocations, the panel recommended use of estimates obtained by averaging the updated county estimates with estimates based on the 1990 census (National Research Council, 1997). For subsequent fiscal years, the allocations were based solely on the updated estimates.

creases in a few states, mostly in the south and the northern mountain states.
- Changes in a state's per pupil expenditure.
- Changes in the amount appropriated for Title I grants. For fiscal year 1999, which controls the allocations for school year 1999-2000, Congress increased the amounts for both basic and concentration grants by about 5 percent.
- State minimums for basic and concentration grants.
- Hold-harmless provisions. As noted above, the statute includes a graduated hold-harmless provision for basic grants, but none for concentration grants. However, for school years 1999-2000 and 2000-2001, Congress established a 100 percent hold-harmless provision for both kinds of grants.

Brown presented the results of an analysis that compared the actual Title I allocations for school year 2000-2001, based on the 100 percent hold-harmless provisions, with what they would have been if only the statutory hold-harmless provisions for basic grants had been applied. Looking at the totals for both kinds of grants, Utah's allocation would have been 21.4 percent less without the 100 percent hold-harmless provisions and the District of Columbia's would have been 8.8 percent more. The 17 states that gained under the 100 percent hold-harmless provisions received in total about $200 million more than they would have under the statutory provisions. Total grants to the remaining 34 states were reduced by the same amount. For the 2000-2001 concentration grants, there was a slight decrease in the total appropriation, with the result that there was not enough money to fully fund the 100 percent hold-harmless provision. Therefore, the Department of Education reduced the previous year's allocation for each district by the same proportion, and new data inputs for that school year had no impact on the allocation.

The second presenter in the session was Graham Kalton, of Westat, who served as chair of the Panel on Estimates of Poverty for Small Geographic Areas. He explained the Panel's mission and described the steps it had taken to evaluate the suitability of the U.S. Census Bureau's estimates for use in the Department of Education's Title I allocations at the county and school district levels. He noted that small area estimates were required for three variables. The key variable is the number of poor school-age children ages 5 to 17. Also needed are the total number of children in that age group and the total population of each area. The total number of children is needed to serve as denominator for an estimate of the proportion of poor children, which enters into the statutory threshold and hold-harmless provisions of the law. Total population is needed in connection with the provision that allows states to use alternate allocations for school districts with less than 20,000 population. Estimates for the first variable are provided by the U.S. Census Bureau's Small Area Income and Poverty Estimates (SAIPE) Program, which was launched in 1992. Estimates of total population and population by age for states and counties come from the U.S. Census Bureau's long-established population estimates program.

The principal sources of data for the SAIPE program estimates are the CPS, the decennial census of population, and administrative record data based on individual income tax returns and participation in the Food Stamp Program. As noted earlier, there are trade-offs among these data sources with respect to timeliness, sampling variability, nonsampling error, and conceptual fit. Therefore, a model-based estimation procedure was adopted to seek the best use of information from each source.

After examining several alternative models for use in county estimates of poor children, the U.S. Census Bureau adopted a log-linear regression model, with the logarithm of the CPS estimate (based on a three-year average to increase sample size) as the dependent variable and data from the decennial census, IRS, and the Food Stamp Program as independent variables. The resulting model was used to predict the number of poor school-age children for each county for 1993 and 1995. These predicted values were averaged with direct CPS estimates for those counties that had actual data from the CPS sample. Within each state, the resulting county estimates were benchmarked to state estimates based on a separate model.

Three methods were used to evaluate the SAIPE Program county estimates. First, standard regression diagnostics for the model selected by the U.S. Census Bureau and several alternative models were examined to determine to what degree the results were consistent with modeling assumptions. Second, the U.S. Census Bureau model and some of the alternative models were used to prepare estimates for 1990, using inputs based on the 1980 census and appropriate data from the CPS, the IRS, and the Food Stamp Program. These estimates were compared with the direct estimates from the 1990 census. Third, for several groups of counties for which the 1993 estimates seemed unusually high or low in relation to previous levels and trends, local officials were contacted to obtain their assessments of the reasonableness of the trends in poverty implied by the SAIPE estimates for 1993.

CASE STUDIES 19

Kalton described some of the results from these evaluations, which were undertaken by joint efforts of the U.S. Census Bureau's SAIPE Program staff and panel members and staff. One analysis by the U.S. Census Bureau showed overestimation of the number of poor children in counties that had a high proportion of the population living in group quarters. This bias was substantially reduced in the revised 1993 estimates by changing one of the independent variables in the model from population under 21 to population under 18. A general conclusion of the panel was that many of the alternative models examined were fairly comparable and that the one selected by the U.S. Census Bureau was a good choice. Nevertheless, the estimates are subject to appreciable error. The comparison of model-based county estimates for 1989 with the decennial census numbers showed an average absolute difference of about 15 percent for numbers of poor children and average absolute proportional differences of 15 percent for numbers of poor children and 16 percent for proportions of poor children. Biases in the estimates for groups of counties with common characteristics are also a concern. In the aggregate, model-based county estimates for 1989, 1993, and 1995, when compared with direct estimates from the CPS, consistently underestimated the number of poor school-age children for the 199 counties with a population of 250,000 or more. Errors in the estimates can also affect the application of the threshold and hold-harmless provisions. An example of this would be a county for which the 90 percent confidence interval for the estimated proportion of poor school age children included 15 percent, the threshold level for receipt of concentration grant funds.

Kalton went on to discuss the estimates for school districts. Several new problems arise at this level: many of the districts are very small, their boundaries change frequently over time, some of them cross county boundaries, and some serve subsets of grades 1 through 12. A key difficulty is that there are no nationally consistent administrative data available to use for the kind of modeling that is done at the county level. The estimation procedure adopted allocated the county model-based estimates for the estimation year (e.g., 1995) in proportion to each district's share of poor school-age children in the 1990 census. For districts whose boundaries changed between 1990 and the estimation year, census data by block were used to obtain census estimates for the districts as defined in the estimation year.

For selected counties, the school district estimation procedure was evaluated in the same way as the procedure for obtaining county estimates (i.e., the procedure was used to estimate the number of poor school-age children in 1989 based on data from the 1980 census and these estimates were compared with the 1990 census results).[4] The average absolute difference in the estimates for poor school-age children was 22 percent, roughly double that which had been observed for the county estimates. The differences in the estimates of the total number of school-age children by school district were smaller, with an average absolute difference of 12 percent, because the allocations were based on the 100 percent data from the previous census and the estimates were compared with 100 percent data from the 1990 census.

Initially the panel was uncertain about whether to recommend use of the model-based county estimates and subsequently had even more difficulty in deciding whether to extend its recommendation to the school district numbers. It recognized the high degree of error in the estimates, but, after examining all reasonable alternatives, concluded that these estimates were at least as good as and probably better than the alternatives.

In concluding, Kalton noted that even though the estimates now being used in the allocation process are more timely than they had been, they still lag the current program year by three or four years. Shifts in the distribution of poverty across the nation in that period of time can be substantial. The levels of variable error and possible persistent error (bias) in the estimates are a subject for concern. In particular, it is important to ask how these errors interact with thresholds and hold-harmless provisions. Finally, he noted that new data will be available soon from the 2000 census and, if it is funded, from the American Community Survey (ACS). These data sources offer possibilities for improving the estimates, but the transition from the current series will raise some complex technical issues.

Bruce Spencer of Northwestern University discussed the presentations by Brown and Kalton. He began by challenging the presupposition that accurate estimates of the number of school-age children in poverty are needed for Title I allocations. A general goal of the program, cited in the legislation, is "distributing resources, in amounts sufficient to make a difference, to areas and schools where needs are greatest" (Improving America's Schools Act of 1994, P.L. 103-382). Another possible goal might be to give

---

[4]This could be done only for school districts that did not change boundaries during the 1980s; also excluded from the evaluation were districts that were not unified, that is, that did not include both elementary and secondary grades.

the federal government some authority over the educational system, which is controlled by state and local governments. It is not obvious why very accurate poverty data are essential to achieve either of these goals. He asked whether congressional intent would be thwarted by the levels of modeling errors in the poverty estimates that were identified by the CNSTAT panel.

Two possible criteria for evaluating improvements are cost effectiveness and fairness (both actual and perceived). If efficiency is the criterion, one needs to ask whether improvements in meeting program goals through the use of better estimates justify the additional costs of producing those estimates. To answer this question would require evaluation of Title I program outcomes for alternative allocation procedures; however, evaluation of Title I under any conditions is notoriously difficult. The concept of perceived fairness suggests that whether or not the allocation formula is optimal in any sense, the estimates used in the process should be as accurate as practicable. However, at some point, resources devoted to making them more accurate may be wasted.

Spencer reviewed some of the sources of error in the SAIPE Program estimates. He echoed Martin David's concerns about definitional problems, census undercoverage, nonresponse, and errors in reporting income, sampling error, and modeling error. He suggested that county estimates for 1986 be compared with the 1990 census data to reflect lack of timeliness. For the school district estimates, he asked whether alternative sources of data, such as participation in the school lunch program, might be used for states for which these data are uniformly available for all districts. Another possibility would be to make direct estimates of the current population ages 5 to 17, using school enrollment data by school district or tax return data geocoded to school districts.

Spencer pointed out that the kinds of data used in formula allocations have many uses and that improvement of these data could yield diverse benefits. It would be a mistake to look only at formula uses of statistics in order to justify their importance or their cost. Other uses may be more important, although harder to identify. For formula-based allocations, the underallocations are offset by the overallocations, so that the loss in overall social welfare resulting from these errors may be a small fraction of the total error.

Another question that needs to be asked is how should the amount of resources spent to obtain the data be influenced by the presence of definitional error? If we have a measure that is invalid for the intended purpose, is it still important to reduce sampling and modeling error? The answer to this question, Spencer said, depends on what loss function is appropriate. If the loss increases moderately with error, less money should be spent on data when the measure is invalid. If the loss from estimation error follows a quadratic function, one should spend the same amount regardless of the size of the bias. Only if it increases even more rapidly with the size of the error, should one spend more on improving the data when the bias is larger.

Henry Aaron of the Brookings Institution began the general discussion that followed by pointing out that the initial allocation is politically driven and that the formula is an instrument for achieving a political outcome. Once a formula is in place, it is very hard to change it. One should focus on how the allocation formula behaves over time. What counts is not whether the formula is accurate, but whether it changes in the same way as the true, target measure.

Several workshop participants called attention to the state per pupil expenditure component of the Title I allocation formula, which had been given only passing attention earlier in the session. This component is defined as current expenditures on public elementary and secondary education (minus certain federal revenue items), divided by attendance as defined by state law. Its value is allowed to vary only within a limited range of the national average. Unlike most states, California until recently defined attendance to include excused absences, which brought down their per pupil expenditure. They have now changed this, which will increase their share of the overall appropriation for the program. One participant, who noted that the inclusion of per pupil expenditure in the formula is commonly justified as a proxy for the cost of educational services, felt that it was not a good measure of relative costs and in fact measures differences among states in the amount and quality of education they want.

There was further discussion of hold-harmless provisions in allocation formulas. There is always a tension between program stability and targeting where the money is supposed to go. The concern of Congress may be less with accuracy in targeting funds and more with the avoidance of sharp differences, sudden changes, and high variability.

Paul Siegel, of the U.S. Census Bureau, had two suggestions about terminology. He objected to use of the term "poverty counts" to describe the SAIPE Program estimates, and he suggested that the term "uncertainty" be used in place of "errors" in the estimates. He also noted that the Congress has provided for updating the SAIPE Program estimates only at two-year intervals, which suggests they may have given some consideration to the costs of producing the data.

THE SAMHSA BLOCK GRANTS

The third case study covered block grants to states for substance abuse and mental health services, administered by the Substance Abuse and Mental Health Services Administration (SAMHSA). Albert Woodward of SAMHSA began the session by describing the current allocation formula and summarized changes in the allocation procedures since the inception of the program in the early 1980s, when several related block grants were consolidated in the budget reconciliation bill for fiscal year 1982 (P.L. 97-35).

Currently, the total annual appropriation for these block grants is about $2 billion, of which about four-fifths goes to substance abuse and the remainder to mental health services. Of the total amount appropriated by Congress each year for the two types of grants, 5 percent goes to SAMHSA for data collection and other administrative activities and 1.5 percent goes to the U.S. territories. The remainder is allocated to the states. The same general allocation formula is used for both the mental health and substance abuse block grants:

$$SALLOC_i = 0.985 \times 0.95 \times AMT \times \left(\frac{P_i C_i I_i}{\sum P_i C_i I_i} \right)$$

$SALLOC_i$ = the allocation to the ith state.
AMT = the appropriation for substance abuse or mental health.
P_i = a proxy measure of the state population at risk.
C_i = cost of services index for the state.
F_i = fiscal capacity index for the state.

The P_i component of the formula, which is a proxy measure of need based on population, is defined differently for the two kinds of grants. For substance abuse, two population groups, ages 18 to 24 and ages 25 to 64, are used, with the urban population ages 18 to 24 receiving twice the weight of the other groups. For mental health services, a weighted average of four age groups of the population age 18 and over is used, using weights specified in legislation.

The cost of services component, C_i, which is the same for both kinds of grants, represents an effort to account for differences between states in the costs of providing services. It is a weighted average of three subindexes covering labor costs, rent, and supplies. In the early 1990s, after SAMHSA was created, the agency used data from the Bureau of Labor Statistics on manufacturing wages by state for the required three-year updates of this component. However, this procedure was challenged by California because the legislation that established the formula had incorporated a research paper that used 1979 census data on wages in nonmanufacturing industries and noted specifically that the Bureau of Labor Statistics data on wages in manufacturing were not appropriate. After consulting with an expert panel, SAMHSA decided to use the most recent decennial census data on wages in nonmanufacturing industries, updated by using wage information for hospitals published annually by the Health Care Financing Administration, an approach consistent with legislative intent.

The fiscal capacity component, F_i, is also the same for both kinds of grants. It is based on the relationship between the Treasury Department's estimates of total taxable resources and the other two components. The larger a state's cost-adjusted taxable resources in relation to its population at risk, the smaller the value of this component in relation to its value for other states.

Woodward reviewed some aspects of the history of these grant programs. In 1981, when the categorical block grants were consolidated, there was a cut of about 20 percent in overall funding and states were given flexibility in the use of the funds received. Allocations were based on total population (as a proxy for need), with no factors representing need or capacity. Legislation passed in 1984 revised the population component of the formula in an attempt to approximate populations at risk and also introduced a factor based on per capita income. As these changes were phased in, Congress introduced a hold-harmless provision. In 1988, there were further changes in the population component of the formula, total taxable resources were substituted for per capita income, and the hold-harmless provision was phased out.

With the creation of SAMHSA in the ADAMHA [Alcohol, Drug Abuse, and Mental Health Administration] Reorganization Act of 1992 (P.L. 102-321), the substance abuse and mental health services block grants were separated and the cost of services component was introduced. As noted above, the change in the labor component from manufacturing to nonmanufacturing in order to conform with the statutory requirement was introduced for the fiscal year 1998 allocations. For fiscal year 1999, Congress introduced a small state minimum, so that no state would get less than 0.375 percent of the total, as well as a hold-harmless provision. For mental health services, every state received at least as much as in 1998 and, for substance abuse, each state got the same amount as for 1998 plus 5 percent.

Woodward drew three conclusions from experience with this formula grant program. First, he said, the formulas represent an attempt to achieve equity among the states. However, congressional changes to the formula, especially through the use of hold-harmless provisions, have interfered with the achievement of this goal. Second, little is known about what impact the federal dollars have had on state spending behavior. A recent U.S. General Accounting Office (1996) report suggests that the federal dollars are supplanting rather than supplementing state funding. It is not clear whether this is what Congress intended. Finally, the current population component of the formula is a very rough proxy for need. The Office of Applied Studies in SAMHSA sponsors a national household survey that collects current data on substance abuse. The National Household Survey of Drug Abuse (NHSDA) has recently been expanded to approximately 70,000 respondents so that it will be possible to produce direct estimates of the prevalence of substance abuse and mental health conditions for eight states and synthetic estimates for the remaining states. These new data may produce better estimates of need at the state level.

The 1992 legislation that created SAMHSA, split the grants between substance abuse and mental health, and introduced several changes into the allocation formula also mandated a study to review the formula and its components and determine whether improvements could be made. The methodology and results of the study (Burnam et al., 1997) were described by John Adams of RAND. His presentation focused primarily on the formula for the substance abuse block grant allocation.

The first step in the study was to examine how the different components of the formula—need, cost of services, and fiscal capacity—interacted. Then, alternative measures of need and costs were developed (the study team concluded that the fiscal capacity measure, total taxable resources, was appropriate and could not readily be improved) and allocations based on these alternative measures were compared with those based on the existing formula. It was not necessarily intended that the alternative measures be readily implementable. The goal was to develop a clearer picture of the main features of the measures that had been written into the legislation. Primary attention was given to the need component of the formula.

The RAND researchers reviewed the legislative history and interviewed congressional staffs in an attempt to clarify program goals. The grants are intended to support both treatment and prevention services related to substance abuse, which in the United States is dominated by alcohol abuse rather than hard-core drug abuse. The nature of the formula, as had been explained earlier by Albert Woodward, suggests a view that the need was greatest among young adults in urban areas.

In an effort to develop a more precise estimator of need, the study team worked with microdata from the NHSDA to develop a logistic regression model to predict the probability of need by state. The dependent variable was an approximation, based on available NHSDA variables, to a generally accepted measure of drug or alcohol dependence. Other NHSDA variables were used as predictors. The model was then applied to data in public-use microdata files from the 1990 census to predict the probability that each person in a state was in need of treatment services.

The usefulness of the NHSDA data was somewhat limited by the fact that RAND did not have access to complete information about the geographic locations of the NHSDA households. Except for a few large metropolitan areas that had been oversampled, the only geographic information available to them was census region and whether or not the household was located in a metropolitan statistical area (urban versus nonurban). Both of these variables turned out to be useful predictors. Contrary to what the framers of the legislation appear to have believed, the estimates of need were generally higher in smaller states with a less urban population. This may reflect the fact that there are more people with alcohol problems than there are with hard-drug problems and that alcoholism is more prevalent in rural areas than some people think.

One very strong predictor of individual treatment needs was gender. However, since the proportion of males does not vary much by state, its usefulness at the state level is limited. Other more useful predictors at the state level were age (highest for young adults), census region (highest probability of dependence in the West), race-ethnicity (highest for white, non-Hispanic), marital status (highest for single or separated), and completed education (highest for high school dropout). For the few large metropolitan areas that had been oversampled, it was possible to use the NHSDA data to make direct estimates of dependence and compare these with the model-based estimates. The agreement was reasonably good for four of the five areas, the exception being Los Angeles.

Some work was also done on the cost-of-services measure. Data from the National Drug and Alcoholism Treatment Unit Survey indicated that labor costs were a somewhat higher proportion of total treatment costs than would be indicated by the weights specified in the legislation. In addition, the survey showed that states spend more to provide services in rural areas, probably because they do not benefit from economies of scale that can be realized in urban areas. The RAND researchers believed that the cost-of-services measure could be made more equitable by adjusting the formula to reflect these differences.

After comparing allocations based on alternative measures with the allocation based on the existing formula, the researchers attempted to devise ways of presenting their findings without calling attention to outcomes for specific states. One global outcomes indicator was the fraction of the total allocation that would be moved. If only the needs component (an estimate of or proxy for the prevalence of drug or alcohol dependence) had been changed, 18 percent of the money would have been moved, and if only the cost-of-services component had been changed, 8 percent would have been moved. If both had been changed, 22 percent of the money would have been moved. A second indicator, which was referred to as "the Senate criterion," was the number of states with changes of more than 20 percent in their shares. A third indicator, "the House criterion," was the fraction of the population living in states with shares that changed by a large amount. Adams presented some tabulations of these indicators. One thing that stood out very clearly was the extent to which using the alternative formulas would cause funds to be transferred from the large, more urban states to the smaller, rural states.

Summarizing what was learned from the RAND study, Adams offered the following conclusions:

- It would be possible to do a better job of defining the population and cost-of-services components of the formula. These improvements would matter, with funds being shifted from larger, urban states to smaller, rural states.
- The formula does not recognize other programs that support substance abuse services. Also, the need estimates are for the entire state population, despite the reality that most block grant funds are spent on services for the poor and the uninsured.
- The statistics used to improve the formula elements would be more useful if there were less clustering in the sample and if more geographic detail could be made available to the researchers.
- The NHSDA, as a household survey, does not cover prisoners and the homeless, population groups that have a higher than average probabil-

ity of dependence. Other data sources, such as emergency room and arrest records, should be sought to fill this gap.

Jane Maxwell, of the Texas Commission on Alcohol and Drug Abuse, was the designated discussant for the session on the SAMHSA block grants. Primarily addressing the population (needs) component of the formula, she cited several reasons why in her view it is not a good indicator of need. Block grant funds are being used to treat medically indigent people, but no measure of poverty is included in the current formula. The allocation formula includes an estimate of the number of persons at risk for substance abuse, but it does not consider what proportion of them are in need of public services. It does not reflect the need for services and it does not take into account different drugs, different populations, or differences among regions. Furthermore, the prevalence of drug abuse is changing. The allocation formula is weighted toward youth, but data from the NHSDA show that the population of drug abusers is aging.

Maxwell presented data by state from two federal data systems that indicate need:

- The Drug Abuse Warning Network (DAWN), which collects data on mentions of drugs in cases from a sample of hospital emergency departments and medical examiners.
- The Treatment Episode Data Set (TEDS), which contains admission and discharge data on all clients entering publicly funded substance abuse treatment programs.

She observed that there was little, if any association between indicators of substance abuse from these datasets and the per capita block grant amounts by state for 1997, which varied from $2.65 for North Dakota to $5.80 for California.

Maxwell concluded her presentation by offering options for improving the need component of the formula. The recent expansion of the NHSDA should provide good data on substance abuse for the eight states that have been oversampled. It is not clear how good the synthetic estimates for the remaining states will be. They should be compared with the results of recent telephone surveys administered by the states. If the agreement is not good, it may be necessary to use data from DAWN or TEDS to adjust the synthetic estimates. Furthermore, it should be understood that household surveys, which do not cover the institutional population, seriously undercount hard-core drug users. She presented data from Texas prison surveys to confirm this point.

In the general discussion that followed, it was noted that the weak association between state per capita grant amounts and measures of need might be explained, at least in part, by the fact that the formula has two other elements—cost of services and fiscal capacity. It was hypothesized that if adjustments were made for state differences in those elements, there might be a better correlation between indicators of need and grant amounts.

Jerry Fastrup provided some elaboration of Congress's reasoning behind the changes over time in the allocation formulas and procedures. He explained that the formula used prior to the 1992 amendments had favored states with a disproportionately high share of the nation's urban population. Responding to the concerns of rural states, the new formula reduced but did not eliminate this differential treatment. At the same time however, the cost-of-services factor, which favored the large urban states, was incorporated into the formula. Also associated with these changes were temporary hold-harmless provisions and features that limited the increase that any state could receive in a single year.

It had been pointed out previously that the SAMHSA block grant formulas did not take into account how much states were spending on substance abuse and mental health programs using other federal funds and their own funds. This appears to be the case for most block grant formula programs. One participant noted that it is difficult to obtain good data on what the states are spending on these programs. For matching grant programs, such as Medicaid, the size of the federal grant does depend in part on what the state spends, but it does not take account of other state programs directed at the same target populations.

Questions were raised about program rules that govern how the states use their block grant funds and about how much information is available about the extent to which program goals were being achieved. One participant observed that information is inadequate to evaluate the relative effectiveness of prevention (including interdiction) and treatment in reducing the prevalence of substance abuse. In the view of some participants, it was not cost-effective to make minor changes in formulas as long as these broader questions were not being addressed. Another view was that the important question whenever a change in a formula is being considered is whether that change is in the right or the wrong direction. The RAND study showed clearly that giving greater weight to the urban population was a move in the wrong direction.

One participant called attention to the contrast between the nature of the formulas used for the SAMHSA block grants and the Title I education grants. For the latter, Congress has left it to an agency, the U.S. Census Bureau, to decide on what data sources and statistical procedures are best suited to produce the estimates needed for the allocation. But for the SAMHSA block grants, nearly all of these details have been specified in the legislation. It was suggested that part of the explanation for this contrast could be that until recently there have been no generally agreed-on measures of the prevalence of substance abuse and mental health treatment needs, and no agency that Congress felt it could rely on to produce such measures. How Congress will react to the improved state estimates to be produced from the NHSDA is not known. Whether they will be used in the formula may depend partly on the expected numbers of winners and losers and partly on whether members of the statistical community can agree among themselves and then convince Congress that these estimates are, in fact, better indicators of need.

THE WIC PROGRAM

The focus of the fourth case study was the Special Supplemental Nutrition Program for Women, Infants, and Children (WIC), a federal grant program administered by the Food and Nutrition Service of the U.S. Department of Agriculture. Ronald Vogel of that agency gave an overview of the program and the procedures for allocating program funds to states and Indian Tribal Organizations. The WIC program provides nutrition and health assistance services to low-income childbearing women, infants, and children. Pregnant and postpartum women and children under the age of 5 who have family incomes not exceeding 185 percent of the applicable poverty guidelines are "income eligible" to participate. To be fully eligible, participants must be determined to be at nutritional risk based on a medical or nutritional risk factor identified by a competent health professional. Persons exhibiting a medical risk factor, such as anemia, underweight, or diabetes, receive higher priority than persons at risk of inadequate nutrition. Participants receive federally prescribed packages of foods designed to meet their specific needs, plus nutrition education and counseling and access to health and social services. The program is currently funded at about $4 billion per year and serves about 7.1 million women, infants, and children each month.

For the first few years of the WIC program, which was established in

1972, state grant amounts were determined at the discretion of the Department of Agriculture. In 1979, a funding formula was established by regulation. There have been several changes in the formula since then, but the underlying objective is to allocate the funds to state agencies in proportion to their share of the national estimate of persons eligible for assistance. The formula does not include a component to adjust for state differentials in the costs of the food packages that are distributed. The state and national estimates are both important: the former are used to determine state shares, and the latter are used by the administration and Congress to determine the total amount requested and appropriated for the program each year. Initially, the estimates were based on decennial census data, but as the program grew there were pressures to produce more timely estimates, leading to the successive introduction of a series of model-based estimates making use of more current survey and administrative data.

Vogel described several issues currently facing the WIC program:

• Citizenship is not a condition of eligibility for WIC program benefits. Failure to include a significant portion of unauthorized immigrants in current population estimates could lower the shares of states with large numbers of eligibles in that category.

• At present, only a few American Indian tribes are participating in the program, but the number is increasing and it may become difficult to obtain the data needed to determine how much of the funding should go to Indian Tribal Organizations.

• There is an "adjunctive eligibility" issue. The current law establishes income eligibility for WIC for persons who participate in other means-tested programs—Medicaid, Food Stamps, and Temporary Assistance for Needy Families—even when their family income exceeds the 185 percent of the poverty cutoff. However, such persons are not included in the estimates.

• Infant formula is a major component of the food packages provided to program participants. Congress has required the Food and Nutrition Service to seek large discounts from formula manufacturers. The WIC program currently pays about $0.35 for a can of formula that would cost from $2.50 to $2.70 in the retail market. Program data indicate that the number of infants currently served by the program exceeds the estimated number of income eligibles by 22 percent, and the manufacturers are concerned about the program's effect on their revenues.

- Like the Title I education and SAMHSA block grant programs, the WIC program includes a hold-harmless provision. Unless the current year's appropriation is smaller than that of the previous year, each state receives at least as much as it did last year, with an adjustment for inflation. When there is an increase in the overall appropriation, the growth funds are allocated to states that have been receiving less than their proportionate shares.

Allen Schirm, of Mathematica Policy Research, described the evolution of the method for estimating the number of eligible infants and children starting with 1979, when the first allocation formula was introduced by regulation. The "first-generation estimator," which was simply the number of eligibles according to the decennial census, was used through fiscal year 1994, based initially on 1980 census data and, for fiscal year 1994 only, on 1990 census data. The shortcomings of this estimator became especially apparent following the 1990-1991 recession. Census and CPS data showed a 20 percent growth in the number of eligible infants and children between 1989 and 1992, and there was clear evidence that this growth had been spread unevenly across states. Thus, the assumption of no change not only created problems for program planning and budgeting, but also led to bias in the proportional allocation of program funds among the states.

The second-generation estimator of WIC program eligibles was an empirical Bayes shrinkage estimator (see Fay and Herriot, 1979; Ghosh and Rao, 1994) that optimally averaged CPS direct sample estimates and predictions from a regression model. The regression model predicted change in the percentage of income eligibles by state between 1989 (the income reference year for the 1990 census) and 1992 (the reference year for the March 1993 Income Supplement) based on observed changes in Food Stamp Program participation, Unemployment Insurance claims, and per capita income. The shrinkage estimates of change in the percentage eligible were added to the estimates for 1989 from the census to obtain state estimates for 1992. These 1992 estimates were multiplied by the U.S. Census Bureau's current population estimates for July 1, 1992, to obtain state estimates of income eligibles. Finally, these state estimates were ratio-adjusted to add to the CPS direct national estimate of income eligibles for 1992.

The resulting estimates were used to determine state food grants for fiscal year 1995. Over $125 million in growth funds was distributed. The estimates were more timely than census estimates and substantially more precise than direct CPS estimates (Schirm, 1995). Had there been no hold-harmless provision in the allocation formula, the WIC grants for about half the states would have differed by at least 10 percent from what they would have been using the first-generation estimates.

It was believed that further improvements could be realized by making use of CPS and administrative data for intervening years and by adding some new components to the regression model. This led to the third-generation estimator, which was used to allocate funds for fiscal year 1996 and has been used with some refinements for all succeeding years. Like its predecessor, the third-generation estimator is an empirical Bayes shrinkage estimator. Inputs for the 1997 estimates, for example, included data for 1989 from the 1990 census, CPS data for each of the years 1990 through 1997, and administrative data for 1989 through 1997. In the past few years, the regression model has predicted the percentage eligible in the estimation year, rather than the change in the percentage eligible, thus eliminating the need to go from levels to changes and then back to levels. Predictors that were added to those used in the earlier model were child poverty rates based on tax returns, tax return nonfiler rates for nonelderly persons, total population, and the "census regression residual," that is, the residual obtained from regressing the WIC eligibility percentage for 1989 on values of the other predictors for that year.

Evaluation of the third-generation estimator showed good relative predictive fit of the regression model according to various criteria and there was no strong evidence of model bias. There have been substantial gains in precision from using CPS data for more than one year.

Schirm concluded his presentation by discussing opportunities for further improvements in the estimates as new and better data become available over the next few years. New data sources will include the 2000 census, the ACS, and possibly the Survey of Income and Program Participation, especially if it becomes a data source for official poverty statistics. In addition, Congress has appropriated funds for increasing sample sizes to improve the precision of CPS state estimates of children lacking health insurance, with the likely effect of also increasing the precision of state estimates of persons eligible for the WIC program. Administrative record data sources will have to be monitored, to identify potential new sources and to look at the data currently in use to see if they remain suitable as changes occur in the programs they are designed to support. Making the transition to new data sources, especially the 2000 census, will raise some complex technical and operational issues. There is still a lag of about three years between the period to which the estimates refer and the period for which funds are being allocated. It may be difficult to reduce this gap, but the possibilities should be examined.

The designated discussant for the session on the WIC program was David Betson, of the University of Notre Dame. He began by observing that case studies are most useful when it is possible to compare and contrast the evidence from more than one case. One issue that is present in most formula funding programs is the availability of relevant data from several different sources, for example income and poverty from the census, the CPS, and the Survey of Income and Program Participation. It is also important to consider how survey data will be used in conjunction with administrative data. The question is how to make optimum use of these varied sources of data.

Betson also noted that another common feature of the case studies presented at the workshop is that significant resources have been devoted to efforts to improve the quality of the estimates used in the funding formulas. He asked to what extent these efforts have been justified by the outcomes, taking into account the existence of program provisions that attenuate the effects of improved estimates.

Betson identified and discussed the implications of another kind of congressionally mandated formula, in the area of child support. In the Family Support Act of 1988, the states were required to develop and adopt mathematical formulas for use in setting child-support awards. Several consequences of this legislation may have some relevance to formula funding programs:

- There was some initial resistance from the judiciary, who did not want to have their discretionary powers usurped by a mathematical formula. But when they were reassured that departures from the formula-based amount would be allowed when justified by special circumstances, they embraced the new procedure as a means of speeding up the disposition of child-support cases on their crowded calendars.
- There was wide variation in the formulas adopted by the states. The clear intent was to develop formulas that reflect the needs of the child in the case being considered, yet only a few states take account of the age of the child.
- Many states define need or ability to pay as 17 percent of the noncustodial parent's income. In practice, this is translated to a nominal dollar amount, which does not automatically change over time, so that no consid-

eration is given to either changes in need as the child grows older or the effects of inflation.

Betson suggested that these observations about child support may be applicable to other formula-based programs. First, even though formula-based approaches may be viewed to take discretion out of the hands of decision makers, the formula can indeed serve their interests. Second, many formula-based programs use only crude proxies for the determination of need. Some judgment is needed to determine whether or not these proxies truly serve the intent of the program. Finally, many formula-based programs adopt ad hoc adjustments to the formula to balance competing needs or demands on the programs. Betson pointed out that the last feature is somewhat analogous to hold-harmless provisions in funding formulas. He observed that downside risks are viewed differently from upside gains but suggested that perhaps, in the interests of stability, some limitations should be placed on the latter as well.

In the general discussion of the WIC program, Schirm was asked to what extent he had examined the consequences of using alternative regression models and predictors. He said that many alternatives had been tried and their properties compared with the ones that were adopted. While others might not have chosen exactly the same estimation procedures, it did not appear to matter very much which of the more promising alternatives was finally chosen.

There was further discussion and clarification of the basis for the hold-harmless provisions of the allocation procedure, their specific nature, and their effects on the resulting allocations. Strictly speaking, it was the Food and Nutrition Service's responsibility, in consultation with the states, to develop the allocation formula and related provisions. However, the agency believed that members of Congress would have had serious concerns if no hold-harmless provisions had been included. At present, the hold-harmless amounts consist of last year's grant increased by some percentage of the accepted rate of inflation, so there is a small amount of money each year for growth allocation purposes, unless there is no increase in funding from the previous year. They have made comparisons of actual allocations with what the state grants would have been if there were no hold-harmless provisions, and there are some substantial differences in both directions. It has been observed that states that are well under their fair share have some of the lowest food package costs.

At the end of the earlier discussion of the SAMHSA block grants, one participant had pointed out the sharp distinction between them and the Title I education allocations. The SAMHSA block grants are based on a congressionally written formula, whereas for Title I, the responsibility for development of the estimation procedures has been delegated to the program agency. A participant noted that the WIC program followed the latter pattern, and he praised the Food and Nutrition Service for supporting the research that has been done in an effort to improve the process. However, in his experience, these two programs—Title I education and WIC—are unique; for most programs the specific details of the allocation formula and process are written into the legislation.

3

Effects on Formula Outputs of Errors in Formula Inputs

As discussed in Chapter 2, several presentations at the workshop focused on what inputs and features (e.g., thresholds) should be included in a particular formula to obtain the desired allocations or what data sources and estimation methods should be used to estimate a formula's inputs. Presenters also discussed the many sources of errors in estimates of formula inputs. These errors include sampling variability, bias, and lack of conceptual fit between the inputs specified by legislation or regulation and the inputs that are estimated from survey or other data.

Although it is widely understood that such errors in formula inputs can lead to errors in formula outputs (i.e., allocations to states or other jurisdictions), it is less well known that the effects of errors in formula inputs can be amplified, attenuated, or influenced in other ways when the errors in inputs interact with the properties of formulas. Thus, one session of the workshop was devoted to formal, statistical issues via presentation and discussion of two papers that explored how estimation errors in formula inputs can affect formula outputs in a single year and over time. This session focused on the interactions among data sources, estimation methods, and formula features and their combined effects on formula outputs, which are—like the formula inputs—statistical quantities. The papers presented showed how such interactions can produce sometimes unanticipated results.

First, Alan Zaslavsky of Harvard Medical School discussed findings from work conducted jointly with Allen Schirm of Mathematica Policy

Research. Zaslavsky began by reviewing the statistical properties of estimation methods and data sources. He contrasted direct and indirect estimates. Direct estimates for a "domain" (defined by both geography and time) are based on data from that domain only. Indirect estimates are constructed using direct information and information from other domains, such as other geographic areas or other time periods. Thus, indirect estimates can be spatially or temporally indirect, or both. Zaslavsky noted that many if not most estimates that have been used for fund allocations are indirect.

Next, Zaslavsky discussed the relative limitations of the principal data sources for estimating formula inputs: the decennial census, current surveys such as the Current Population Survey (CPS) and the Survey of Income and Program Participation, and administrative records. He then described the potential implications of introducing the American Community Survey (ACS) as a source of data for allocating funds. His main point was that introducing a new data source could have substantial consequences for allocations because it changes the estimation errors in formula inputs and may change the frequency at which allocations are recomputed.

Zaslavsky concluded his review of formulas and their inputs by discussing some common formula features. Formulas often contain features that cause allocations to be disproportionate to need even though proportional allocation may be the primary objective. Such features include hold-harmless provisions that limit downward fluctuations in funding and thresholds that require a minimum level of need for distribution of funds, thus concentrating funding where it is most needed.[1]

In the remainder of his presentation, Zaslavsky focused on the interactions between the statistical properties of data sources, estimation methods, and the resulting estimates of formula inputs and the features of funding formulas. He began his discussion of these interactions by stating several general results:

• In the fund allocation process, the procedure for estimating formula inputs cannot be separated entirely from the funding formula. For example, a formula that specifies the use of a moving average of estimates

---

[1] Some threshold provisions are designed to avoid distribution of amounts so small that they could not be used effectively by the areas receiving them. Others are designed to channel resources to those areas whose needs are greatest in absolute or relative terms.

for three single years produces the same allocations as a formula that specifies the use of the "best" estimate for a year if that estimate is technically implemented as the moving average over three years.

- Two main implications of using decennial census estimates of formula inputs are that allocations are stable in most years with possibly large shifts every 10 years (depending on what hold-harmless provision may pertain) and that allocations are sensitive to the particular socioeconomic and demographic conditions in the census reference year rather than to the average conditions over a decade.
- The effects of a hold-harmless provision depend on the frequency with which fund allocations are recomputed.
- Averaging over time reduces the variances of estimates of formula inputs.
- If the estimation procedure and the funding formula are linear, allocations will be unbiased, that is, correct on average over time.

After describing these relatively straightforward, general results, Zaslavsky presented simulation results that illustrate the more complex interactions between the statistical properties of estimates and the features of allocation formulas. The simulation scenarios were defined by those statistical properties (e.g., method of estimation and magnitude of sampling error) and formula features (e.g., presence of a hold-harmless provision or a threshold). Amounts allocated to each geographic area were determined independently.[2]

Principal findings from these simulations were:

- When there is a threshold in a formula, sampling variability in estimates of formula inputs smooths allocations toward the threshold, an effect that is strongest for areas whose true need is near the threshold. As sampling variability rises, areas whose true need is below the threshold for receiving funds are more likely to receive funds, and those whose true need is above the threshold are more likely to receive nothing. On average, areas with true need below the threshold get more than they deserve, while areas with true need above the threshold get less than they deserve. The amount of smoothing of average allocations toward the threshold increases as sam-

---

[2]For a full description of the specifications for these simulations and the results, see Zaslavsky and Schirm, 2000.

pling variability increases. Thus, there is a tendency for the allocations for smaller areas (which typically have smaller samples and larger sampling errors) to be distorted more than the allocations for larger areas. This implies that the sampling plan for the data source used to produce estimates can affect the allocation of funds, an effect that is almost certainly not anticipated when statisticians specify the sampling plan or when policy makers specify a threshold for a formula.

- When there is a hold-harmless provision in a formula that allows an area's allocation to rise by any amount but fall by only a limited amount, sampling variability in estimates of formula inputs "ratchets up" allocations over time. The amount of ratcheting increases as sampling variability increases. Thus, smaller areas tend to benefit more from a hold-harmless provision than larger areas because the upward bias in allocations is greater for the smaller areas.

- Using moving average estimates (e.g., averaging information from the most recent three years) can greatly reduce the biasing effect of a hold-harmless provision.

Next, Zaslavsky discussed the implications of assuming statistical independence of the geographic areas. He noted that for most formula allocation programs, this specification is not strictly true because the sum of the amounts allocated for each time period must equal the total amount appropriated for the program. When total funding is fixed, an undeservedly high allocation to one area (due to a fortuitous sampling error) comes at the expense of areas that were not so fortunate, so these other areas are allocated less than they would have been if total program funding were open-ended. Nevertheless, based on algebraic derivations, Zaslavsky argued that under some—but not all—circumstances, the relative biases in allocations to different areas under such fixed funding are essentially the same as the relative biases found in the simulations under the assumption of open-ended funding. One circumstance under which biases might be notably different with fixed funding is when a small number of areas substantially influence the estimate of the basic allocation parameter (e.g., the dollars allocated per eligible person).

Zaslavsky concluded his presentation by observing that there are inherent conflicts of values in allocating funds. If allocations are responsive to changes in need, an area's funding may be unstable, whereas if funding is stable, it may not be responsive to changes in need. When policy makers specify a formula that is responsive to increases in need and impose a hold-harmless provision to ensure stability (in the form of protection against a substantial drop in funding), allocations will be biased. The biases will be different across areas in ways that are unrelated to differences in need. Some of these conflicts in values could be lessened if all participants in the allocation process worked together and better appreciated that data sources, estimation methods, and the funding formula are all part of a single process. Policy makers could then take account of the properties of potential data sources and estimation methods when designing formulas, and statisticians could take account of policy objectives and formula properties when evaluating new data sources or estimation methods.

David Betson of the University of Notre Dame gave the second presentation. Betson first addressed whether the estimated biases reported in the paper by Zaslavsky and Schirm are sensitive to the specification of open-ended funding used in their simulations. To investigate this issue, Betson conducted simulations under both fixed and open-ended funding. He discussed the following findings:

- In the absence of hold-harmless provisions and thresholds, there are no biases in allocations if funding is open-ended. However, if total funding is fixed, there may be small biases. A few of the largest areas (those with stable estimates) may receive slightly less than they would receive under open-ended funding, while some of the remaining areas may receive slightly more.
- In the presence of a threshold, the biases in allocation under open-ended and fixed total funding are about the same except for large areas with true need substantially above the threshold. Under open-ended funding, allocations to such areas are approximately unbiased, while under fixed funding they may be somewhat downwardly biased. However, this finding pertains only when a substantial fraction of the total true need is in areas with true need near the threshold.

Next, Betson addressed the goal of stable funding for every area. Stability is often justified on grounds of equity, with the claim that large decreases in funding are unfair. Betson proposed that stable, predictable funding is also needed if funds are to be spent effectively. Money may be wasted if there are large swings in funding, whether they are up or down. Two methods of achieving stability are to include a hold-harmless provision in a formula or to smooth estimates of formula inputs by calculating moving averages. Alternatively, the allocation formula can be smoothed in a way that is not explicitly temporal. For example, Betson noted that stability in funding could be increased if the step function that defines a funding threshold is replaced by a logistic function or other smooth function. The limitation of this approach is that there is a tradeoff between the goal of stability and the goal of concentrating funds where need is greatest. As the allocation formula becomes smoother, there is less concentration of funding. Thus, as noted earlier in the presentation by Alan Zaslavsky, attractive policy objectives are often in conflict in practice, even when policy makers agree on the objectives.

Paul Siegel of the U.S. Census Bureau led off the discussion of these two presentations. He began by observing that allocations are the result of combining a formula and estimates of the formula's inputs, and those estimates are the result of combining truth and statistical error. Generally, allocations based on estimates will not be the same as allocations based on truth. The Zaslavsky-Schirm and Betson papers demonstrate that the differences between these allocations are exacerbated by features of the formulas. But, according to Siegel, that may simply reflect the fact that formulas result from many inevitable compromises over many different goals. Those compromises and perhaps some of the biases that may result from them are not necessarily inefficiencies or deficiencies that must be eliminated.

However, Siegel was troubled by the fact that, as demonstrated by Zaslavsky and Schirm, areas with the same true need could receive different expected allocations simply because need was measured with varying precision. This finding suggested to him that one important advantage of using model-based estimates, such as the estimates from the U.S. Census Bureau's Small Area Income and Poverty Estimates (SAIPE) Program, is that the model-based estimates are more precise and are much more equal in precision across areas than are direct estimates.

Siegel concluded by remarking that the paper by Zaslavsky and Schirm has illustrated many of the potentially troubling ways in which properties of estimates and features of formulas interact. He noted that Betson has shown that these interaction effects may be significant in some areas and negligible in others. Thus, it is important to begin assessing how well actual allocation processes are performing.

Robin Fisher of the U.S. Census Bureau was the second discussant. He showed graphically how, in the presence of a threshold, an area's expected allocation could change as the area's true need changes. Fisher noted a problem that Zaslavsky and Schirm found in their simulations: when an area's true need is not far above the threshold, the area may receive much less than it deserves on average. This bias occurs because when there is substantial variability in estimated need, there is a substantial probability that the area's estimated need falls below the threshold and the area receives no funds.

Fisher noted that he interprets a formula as an expression of policy makers' intent, although some of the previous discussion at the workshop had led him to question that interpretation. If a formula truly reflects such intent, abstract arguments to make formulas smoother—despite their appeal to statisticians—might not convince policy makers. However, illustrating how very small errors in estimates can cause some areas to lose all of their funding may persuade policy makers that some changes to formulas could enhance fairness.

The remarks by Siegel and Fisher were followed by a period of open discussion. One workshop participant asked Alan Zaslavsky what specific suggestions he would make for replacing a step function in a funding formula. Zaslavsky responded by noting that there are two problems with step functions.[3] First, they can produce allocations that are highly variable. Second, the allocations, on average, depend on the sampling properties of estimates of formula inputs. Surely, neither of these outcomes were intended by policy makers when they specified a step function. Thus, it may not be hard to convince policy makers that a smoother function would be desirable. One drawback to a smoother function that would have to be addressed is that some of the amounts allocated will be smaller than the amount implied by the threshold of the original step function. Thus, one would need to consider the specific program to determine whether such amounts would be too small to be effectively spent. Zaslavsky observed that piecewise linear functions like those that Congress has specified in the tax code would be simpler and easier to explain than the logistic function (discussed by David Betson) but have a similar effect. Betson added that

---

[3] A threshold, in which nothing is received if the estimate of need is below a specified number or proportion, is one form of step function. Another example occurred in the Title I education allocations for fiscal year 1997, in which the hold-harmless provision applied to basic grants at variable rates. Counties and school districts with 30 percent or more poor school-age children were guaranteed at least 95 percent of the previous year's grant. The guarantee dropped to 90 percent for areas with 15-30 percent poor school-age children and 85 percent for areas with fewer than 15 percent poor school-age children (National Research Council, 1997:50).

although smoother functions will reduce some of the problems under discussion, they will not eliminate the problems if they are nonlinear.

One workshop attendee raised the potentially important issue that improving estimates of a formula input may not be all that helpful if the input is only weakly associated with objective measures of program success. For example, obtaining better estimates of poor children for Title I allocations may not improve the overall effectiveness of Title I funds in improving educational outcomes for the target population. Alan Zaslavsky responded to this point, noting that even if the input to a formula were exactly the sole measure of program success and there were no conceptual problems in measuring the input from available data, the problem of sampling error in estimating the input would still exist. Even in this best-case situation, there would still be the unintended effects of interactions between properties of estimates and features of allocation formulas. Many of the approaches to reducing those effects would cost very little and would surely be worth pursuing. A key point of Zaslavsky's response was that the issues raised by the presenters in this session are relevant to real fund allocation processes, even though there may be other important concerns, such as whether a formula's inputs, even if they could be perfectly measured, are appropriate.

4

Roundtable and Concluding Sessions

The penultimate session of the workshop was a roundtable with presentations by four speakers, followed by open discussion. The speakers had been asked in advance to discuss one or more of the following questions:

- Are there problems with the quality or timeliness of available data?
- Are there features of new or future datasets that are particularly relevant to issues of formula allocation?
- Do you think that the estimates used in formulas or the features of formulas have unintended consequences with respect to equity between jurisdictions? If so, what changes might resolve such problems?
- Do you have any suggestions for changing formulas, data, and estimation procedures?
- What issues could be usefully addressed by the Committee on National Statistics in a study of statistical and data needs for allocation formulas?

Paula Schneider, of the U.S. Census Bureau, led off by discussing the use of the U.S. Census Bureau's data products in funding formulas. She focused on potential future uses of data from the American Community Survey (ACS), which is now being developed. The ACS has been designed as a continuing household survey which, when fully implemented, will provide annually updated demographic and economic information for small areas. Content will be similar to that of the decennial census long form. Testing for the ACS started in 1996. In April 2000 the survey was operating at 31 sites across the country, with sufficient sample size to produce estimates down to the census tract level, so it will be possible to compare these estimates with data from the 2000 census long form. Assuming that Congress provides funding, the ACS will become fully operational in 2003, with estimates for areas as small as 65,000 population available in 2004 and updated annually. By 2010 there will be no need for a long form in the decennial census.

As part of the 2000 census evaluation program, an ACS questionnaire was sent to a national sample of about 700,000 households. State estimates based on this supplementary survey will be available in July 2001. These data can be compared with the long form data and could be used in conjunction with selected funding formulas to see what would happen if the new data were used.

Schneider emphasized that it will be up to Congress, working with program agencies and the U.S. General Accounting Office, to specify what data should be used in funding formulas. She believed that data from the ACS, supplemented by data from other surveys and administrative records, have the potential to improve the timeliness and quality of estimates used in funding formulas. She gave two examples:

• Grants to states under the Individuals with Disabilities Education Act. At present, the state shares are still being determined on the basis of data from the 1990 census long form.
• Community development block grants that go to metropolitan areas. ACS data for the largest areas will be released starting in July 2004, to be followed by data for successively smaller areas over the next four years.

Linda Gage, the state demographer with the California Department of Finance, discussed the role of current population estimates in determining the relative amounts received by states under various funding formulas. For example, the U.S. Census Bureau's estimates of total population are used as control totals for Current Population Survey (CPS) estimates and for the estimates of school-age children in poverty that are produced by its Small Area Income and Poverty Estimates (SAIPE) Program for use in the allocation of Title I education grant funds. In the future, they will be used in the same way to produce estimates from the ACS. She noted the importance of the issue of whether all of these estimates should be adjusted for the census undercount. A 1999 U.S. General Accounting Office study showed that if adjusted 1990 census numbers had been used in fiscal year

1998, California would have received an additional $223 million in formula grant funding. If the adjusted 1990 census numbers had been used throughout the decade, the state might have received an additional $2.2 billion. A recently released study by PricewaterhouseCoopers estimated that if unadjusted rather than adjusted numbers from the 2000 census are used, California will lose an estimated $5 billion in the years 2002 to 2012.

Some states, including California, have their own population estimates programs used for allocating grant funds within their states. During the 1980s, the U.S. Census Bureau averaged its estimates for California with the state's independent estimates, which were higher, and published these averages as their official estimates. The 1990 census count for California exceeded both the U.S. Census Bureau and the state estimates. After the 1990 census, both the U.S. Census Bureau and the state of California evaluated their estimation procedures and introduced changes. The two sets of estimates continued to diverge and currently the state's estimate exceeds that of the U.S. Census Bureau by about 900,000 persons. Taking into account this difference plus the 1990 census estimate of undercount for California, Gage believes that the numbers currently being used in funding formulas are about 5 to 6 percent below the state's true population. She expressed her concern about the effects of differential underestimation on funding to the states, and her hope that evaluation of estimation procedures following the 2000 census can lead to changes that will reduce the extent of the divergence between their estimates and those produced by the U.S. Census Bureau.

The third roundtable speaker was Katherine Wallman, chief statistician, U.S. Office of Management and Budget. She is responsible for the development and periodic updating of several classification systems used in the collection and presentation of official statistics, covering concepts such as industry, occupation, race and ethnicity, poverty, and metropolitan areas. These classification systems are developed solely for statistical purposes; programmatic uses, such as regulation or allocation of funds, do not influence their general structure or specifics. Nevertheless, in practice they are often used to determine eligibility for federal assistance or to allocate funds to eligible areas. She provided several examples of how the metropolitan-area classifications have been used in these ways. For example:

• Under the legislation that governs the Medicare program, reimbursement rates for hospitals vary significantly, depending on whether or not they are located within metropolitan areas.

- Under the Rural Revitalization Through Forestry Program, the term "rural community" means any county that is not contained within a metropolitan area as defined by the U.S. Office of Management and Budget.
- In the Urban Park and Recreation Recovery Program, the secretary of the interior is authorized to establish eligibility to general-purpose local governments in standard metropolitan statistical areas.
- Under the Rural Homelessness Grant Program, which makes grants to organizations providing direct emergency assistance to homeless individuals and families in rural areas, the terms "rural area" and "rural community" mean any area or community, no part of which is within an area designated as a standard metropolitan statistical area by the U.S. Office of Management and Budget.

Other examples covered such diverse areas as mortgage insurance, tax credits for low-income housing, organ transplants, and immigrant visas. In a recent review of the U.S. Code, her office found at least 20 such references to the metropolitan-area construct.

John Rolph, of the University of Southern California and chair of the Committee on National Statistics, was the final speaker. He reviewed some of the earlier discussion of the effects of thresholds and hold-harmless provisions used in fund allocation formulas. Clearly, it is not always well understood in advance how the formulas will operate, and they do sometimes have unintended effects. As noted by several speakers, there are important conceptual and measurement error questions associated with data elements included in funding formulas. Two fundamental questions need to be addressed: (1) What might be done to mitigate the effects of errors or uncertainty in the formulas? (2) Is correcting these flaws of principal importance or are there more fundamental questions that need to be addressed about how formula allocation processes operate?

Commenting on the first question, Rolph noted that some ways to improve formulas had been suggested, including replacing thresholds by S-shaped, continuous functions and using moving averages rather than hold-harmless provisions to dampen the effects of large year-to-year changes. He suggested that better correspondence between the intentions of those who draft legislation and the actual formula might be achieved by creating an analytical resource, perhaps located in the U.S. General Accounting Office or the Congressional Research Service, that would provide real-time advice and appropriate modeling and simulation capabilities to legislative staff.

With regard to what may be more fundamental issues, Rolph noted that the Committee on National Statistics had been exploring the possibility of undertaking a panel study on formula allocation. Such a panel would be a logical follow-on to the Panel on Estimates of Poverty for Small Geographic Areas, to this workshop, and to the panel that was being established to evaluate the estimates used in the WIC program. He asked workshop participants for suggestions as to what issues should be on the agenda for the proposed panel.

In the general discussion that followed, several themes were suggested for consideration by the proposed panel on formula allocations. Some speakers proposed that the panel should go beyond an examination of the formulas themselves and how they are affected by the quality of the statistical data used as inputs. It should study the conceptual foundations of grant formulas and the desired outcomes with regard to efficiency and equity. It should address questions such as how to integrate measures of need, cost, and fiscal capacity into a formula and whether there are other components, such as outcome measures, that should be included.

Some participants alluded to the resources provided for research on improving the data and the formula allocation procedures used in the Title I education and WIC programs, and suggested that other formula allocation programs might benefit from similar provisions. However, one participant cautioned that not all program agencies have the same ability to sponsor and monitor relevant research and to apply research findings to the allocation process. Others pointed out that there is a limit to how much accuracy is needed; there is some point beyond which the costs of further improvements outweigh the benefits from any gains in efficiency and equity. Some felt that more resources should be devoted to the evaluation of program outcomes. One attendee, referring back to the presentation by Linda Gage, reminded the group that current population estimates will continue to play an important role in providing updated estimates for use in allocation formulas.

Finally, one person urged careful consideration by the panel of the intended audience for its findings. If the intention of the panel is to affect the process of writing formulas for the allocation of funds, its reports should be prepared with the idea that they are directed primarily at Congress and the agencies that have been assisting Congress in this process.

Henry Aaron of the Brookings Institution wrapped up the workshop. He started by noting that there has been considerable dedicated and talented work to produce improved measurements for use in formula allocation. However, in his view, relative to other kinds of needed work, such as exploring the effects of varying the program rules, the marginal value of additional efforts to improve the quality of input data is relatively small.

He then asked what role formula allocation plays in the political process, considering that it would be possible for Congress to vote direct dollar amounts and that it sometimes does so. He answered this question by noting that allocation formulas are a device to achieve closure to what would otherwise be unendable debates. Members of Congress are elected by constituents to serve their interests. If they do not strive to get the largest possible appropriations for their states and districts, they will pay for it later. However, they can shield themselves from unpleasant consequences if they can point to a plausibly objective formula and say they did the best they could, but this is what the formula produced.

To support his subsequent remarks about formula allocation programs, Aaron defined the following concepts:

G = the goal(s) of a federal grant program, e.g., educational outcomes.
T = a government transfer of funds.
T_i = transfer to the ith jurisdiction.
I_i = an indicator of true need for the ith jurisdiction.
e_i = the difference between T_i and I_i.
P = a politically determined goal for the transfer.
R = program rules to determine resource allocations within districts.

The error term has two parts. It consists of any conceptual error separating the transfer amount from the (unknowable) indicator of true need and errors in measurement or estimation that distinguish the numbers actually used from ones that are potentially more accurate.

Aaron emphasized that it was important to make the distinction between T and R. The former is simply a transfer of money but the program rules determine what goes on in each jurisdiction when the money is received. Transfers of funds can affect recipients in ways that are not necessarily intended or obvious. For example, in the legislation under which Medicare distributes extra funds to hospitals that take care of disproportionate numbers of low-income persons, the amounts hospitals receive are determined by the numbers of Medicare and Medicaid patients they serve. This measure of need provides an incentive for states to have a broad but shallow Medicaid program that covers as many people as possible. Not surprisingly, the legislation that established this program was initiated and supported by the Senate Finance Committee at a time when Russell Long, whose state of Louisiana had such a program, was chair of the committee.

It is also important, Aaron said, to look at the effects of grant funding on overall government spending. By giving money, you may change incentives within a state. Under a matching grant program, you can bring in additional funds. If the expenditures are in a new area, political constituencies may arise on behalf of activities that never before existed. But state legislators might find funds for these new activities by reducing appropriations for other programs. There is a case for a quasi-anthropological approach that looks very carefully at the institutions and details associated with particular grant programs.

Aaron said he believed that, for two reasons, the impact of mistakes on the achievement of program goals is typically very small. First, suppose we start from an allocation where marginal dollars spent yield equal marginal benefits to all jurisdictions. If we make a mistake, the marginal benefit to the jurisdiction that receives more money will decline a bit and for the jurisdiction that receives less money it will increase slightly, but the impact of these changes on welfare is distinctly second order. Second, evaluation studies suggest that some allocation programs are not very effective in achieving program goals. For example, evidence that the Title I education program has had significant effects on educational outcomes is very limited.

In his judgment, Aaron said, the process of collecting data for use in allocation formulas on topics, such as poor children, nutritional intake, and various illnesses, has an important educational function in shaping political views about what is considered decent and acceptable. Nevertheless, in the final analysis, to some degree the choice of formula inputs reflects a political compromise or consensus, and this is what determines the amounts of money transferred. The effects of these choices over time are difficult to predict.

He stressed the importance of program rules. In discussing the Title I education program, the workshop participants reviewed the procedures for allocations to each state, county, and school district, but that is where the analysis ended. He suggested that what happens after the funds get to the school district is an order of magnitude more important than marginal adjustments in the distribution formulas. Should poor children receive special instruction outside the classroom? If a poor child transfers to a magnet school, should Title I funds also be transferred to that school? Answers to questions like these could have a major effect on the evolution of the educational system and its efforts to meet the special needs of poor children. It is important to pay attention to formulas and their inputs to guard against the possibility that some groups may try to manipulate the process to get grossly more than their fair share. What is more important, however, is to understand what makes a program work. Aaron observed that statisticians, working to address this question in collaboration with economists, sociologists, and other social scientists, can make important contributions.

PART II

Panel Report

Following the April 2000 Workshop on Formulas for Allocating Program Funds, described in Part I of this report, the Panel on Formula Allocations was established within the Committee on National Statistics to conduct a study of formulas used to allocate federal and state funds.[1] The study focuses on the statistical estimates used as inputs to formulas, data and methods for estimating these inputs, the features of the formulas, and how estimates and formula features interact in ways that affect outcomes. The panel study is considering in greater breadth and depth how the properties of estimates, such as sampling error and response quality, can affect their use in formulas that have a variety of features (e.g., thresholds for eligibility and hold-harmless provisions).

The purpose of the study is to provide a detailed assessment and several illustrations of how formula features can interact with estimator properties in ways that affect the likelihood of program goals being met; it is not designed to recommend changes in existing formulas, new formulas, or the use of particular datasets. The work of the panel will include analyses of allocations with a variety of program provisions and estimators for programs that cover a range of areas, such as education, community development, public health, and others.

---

[1]Formulas play a central role in cost of living escalator clauses, labor contracts, child support rules, income tax collection, and congressional apportionment. Although these can serve as interesting and relevant analogies, we do not consider them in this report.

In this initial report, the panel summarizes themes identified in the workshop and in its first three meetings, highlights the principal issues it intends to address, and outlines anticipated activities. Themes and issues are presented in Chapter 5 and anticipated activities are described in Chapter 6.

5

Themes and Issues

THE FORMULA ALLOCATION PROCESS

Programs that allocate federal funds to states and localities address three principal goals: delivering funds to the right places, implementing programs and delivering services, and producing the desired outcomes (e.g., health improvement, educational attainment). Formulas have a major, direct role in achieving the first goal; a substantially smaller, indirect role in achieving the second goal; and essentially no role in achieving the third goal, except through the first two goals.

There are three models for the way that Congress interacts with Executive Branch agencies in designing programs to allocate federal funds: (1) Congress can legislate the exact amount to be received by each state or other political subdivision; (2) Congress can pass legislation that includes a detailed formula for determining allocations; or (3) the legislation can describe program objectives and leave it to the Executive Branch agency to determine the formula or other process for allocation. Occasionally, an agency receives funds for research on improving the formula inputs.

An example of the first approach is the Capitalization Grants for State Revolving Funds Program of the Environmental Protection Agency; the Medicaid program is an example of the second. The Special Supplemental Nutrition Program for Women, Infants, and Children (WIC) and Title I of the Elementary and Secondary Education Act provide two examples of the third approach. In recent years, the U.S. Department of Education has improved the accuracy and timeliness of the estimated numbers of school-age children in poverty for Title I allocations (via the U.S. Census Bureau's Small-Area Income and Poverty Estimates Program [SAIPE]). However, the programmatic effects of these improvements have been largely negated by legislated hold-harmless provisions. Such post hoc modifications may serve short-term political goals. Nevertheless, when total program funding is approximately level, avoidance of funding reductions to some political subdivisions can be accomplished only at the expense of other jurisdictions that would have received additional funds.

When a program is authorized (or reauthorized), the explicit nature of a formula facilitates the legislative process and increases actual and perceived fairness. Scenarios can be evaluated and the formula specification adjusted. A formula can make the allocation process more transparent and promote full disclosure. Once implemented, a formula may be fine-tuned or considerably modified. Communicating the rationale for and evaluating an allocation process without a formula is far more difficult.

In the evaluation of programs and allocations, technical statistical issues are important, but so are many other administrative and political aspects of a program. If a program fails to attain its stated goals, the fault may lie in requirements dictated by the legislation, in the data inputs to the allocation formula, in the allocation process, or in the program services. There is a complicated interaction between formula inputs and features, with the possibility for unintended and unanticipated, cascading consequences of various combinations. For example, hold-harmless provisions attempt to balance the goals of changing allocations when necessary and maintaining funding stability. As discussed in Part I, Chapter 3, hold-harmless provisions may systematically and persistently help or harm areas based on the statistical properties of formula inputs rather than changes in true need. The absolute and relative impacts of hold-harmless provisions depend on whether total funds for a program are capped or can adjust to the formula-based financial need.

DESIGN AND EVALUATION OF FORMULA ALLOCATION PROGRAMS

As the foregoing indicates, design, implementation, and evaluation of a funds allocation program is considerably more complicated than a superficial view would suggest, entailing synthesis of statistical information from several sources and complex processing of the results. Many programs use formulas with several elements, such as need, capacity, and cost of services. Furthermore, as pointed out by Zaslavsky and Schirm (2000), there is a degree of arbitrariness in what constitutes the inputs and what constitutes the formula. Legislation can direct that "appropriate" statistical analyses be used to prepare inputs (as is the case for Title I Education) or such analyses can be built into the formula (as is the case for Medicaid). For example, the law might specify that allocations are to be based on a three-year moving average and that each year's estimate is to be based on a single year's data. However, the same effect would be obtained if the formula called for an estimate for a single year but, based on a statistical assessment, the estimate for that year was calculated as a three-year moving average.

This complexity calls for a framework for evaluating program performance. It should include measures of monetary allocation success (possibly including loss functions that compute equity/inequity measures), effective use of funds for program-specified services, and beneficial impacts of services, and it should be used to assess performance of current programs and to recommend changes. Developing such a framework will be challenging. Challenges include defining equity, especially when formulas include multiple elements of need, fiscal capacity, and effort, and balancing equity with efficiency and political considerations. Although evaluation of an individual allocation program is of primary importance, evaluating how different federal allocation programs interact and how they impact programs that allocate state funds for similar purposes is also important.

During the legislative process, the U.S. General Accounting Office and the Congressional Research Service, when requested by congressional staffs, provide simulations that evaluate formula options. There are institutional mechanisms for obtaining additional expert input during the development, evaluation, and modification of allocation formulas. In those instances when it is left to an Executive Branch agency to determine the formula or other process for allocation, the agency will need similar capacity to implement and evaluate formulas and to work through the public comment process. State agencies have similar needs, both in developing their own allocation formulas and in understanding how the federal formula allocation programs work. The parties to these processes do interact, sharing problems and solutions, but additional communication and coordination could be beneficial.

Many take for granted that efforts to improve the quality of formula inputs (accuracy, conceptual relevance, timeliness) are desirable. But the benefits may not justify the costs. For example, the U.S. Census Bureau's small-area income and poverty estimates are updated on a two-year schedule. The cost of annual updating may not be justified by the consequent improved performance; such trade-offs need to be evaluated. To get started, one can consider the situation wherein there is no sampling error and the model is correct and evaluate the improvement relative to the current situation. Performance measures will be needed to conduct such an evaluation.

DATA SOURCES

A few key data sources such as the decennial census, the Current Population Survey, and Internal Revenue Service records are widely used to support formula allocation programs. In addition to these data sources currently in use, new information sources have the potential to improve formula inputs. The American Community Survey (ACS), which is intended to replace the decennial census long form, would be a major new data source that might be used in estimating inputs if the survey is implemented as planned. With data from census 2000 becoming available in stages and the ACS pending, an immediate and high priority should be given to developing recommendations on how to make a smooth transition to these and other data sources and how to evaluate the impact on allocations of introducing new data sources.

Guidance is also needed on the statistical limits of available information. No formula will work perfectly, especially at a fine level of spatial, temporal, and demographic disaggregation. Inevitably, there will be short-term statistical fluctuations in estimates used as formula inputs. While reduction of these statistical fluctuations is an important objective, considerable attention must be paid to identifying and reducing persistent biases for identifiable geographic and demographic subgroups to keep allocations aligned with true needs (see National Research Council, 2000b, for examples).

In some circumstances a lower limit to the population size of areas for which the federal agencies are required to produce population and income estimates has been proposed. For example, the 1980 Panel on Small-Area Estimates of Population and Income recommended that the U.S. Census Bureau not provide postcensal population estimates for places with population below a (to be determined) threshold (National Research Council, 1980). However, the Title I education program now mandates allocations to school districts, and many school districts have small populations. Since some process (formal or informal, sophisticated or naive) will be used to determine school district allocations, it may be preferable to base them on a standardized approach using the best available information and statistical analysis. Properties of the estimates and their effect on allocation programs should be assessed and, in the spirit of openness, the estimates should be made public.

Administrative data are an important element in developing model-based statistical estimates used as formula inputs (for example, Internal Revenue Service and food stamp records are used in the U.S. Census Bureau's SAIPE Program). Evaluation of the potential for modifications in the collection and storage of program and administrative data to improve their performance as inputs to allocation formulas will be beneficial. For examples of possible modifications, see National Research Council (2000a:Ch.5).

THE ROLE OF THE STATES

While policy makers, program administrators, statisticians, and others at the federal level give their attention to formulas and formula inputs, some recipients of federal funds, especially states, are spending significant resources on efforts to increase their allocations. As discussed in Chapter 2, California changed its definition of school attendance to achieve a per pupil expenditure estimate that increased its share of Title I education funds, and the state protested as improper the use of manufacturing wages in the mental health/substance abuse block grants. These examples show that states will challenge definitions and data sources to maintain or increase their allocations. It is therefore important to document what individual states are doing to improve their shares of formula funds and to consider what federal agencies can do to ensure a level playing field.

Further consideration of the role of states in fund allocation processes suggests that adjuncts to the use of national databases should be considered for developing formula inputs. For example, a program could require that a state or county produce "best estimates" and use these in a formula. This approach has the potential to produce more timely and targeted estimates. In fact, many funds allocation programs currently allow flexibility in state-provided information, especially for distributing federally allocated funds within the state. If similar flexibility is allowed in providing inputs to the formula for allocating funds *among* states, inputs should be as immune as possible from manipulation of definitions or data sources.

Formulas also figure importantly in the allocation of state funds to counties, cities, school districts, and other jurisdictions. State funding of elementary and secondary education substantially exceeds the contribution from the Title I and other federal education programs. Since the early 1970s, perceived disparities between school districts have led to many efforts, in state legislatures and in the courts, to develop more equitable allocation processes (National Research Council, 1999b). A recent court decision by a New York State judge declared the state's method of financing public schools illegal and set a September 2001 deadline for the state to revise its formula (Goodnough, 2001).

6

Anticipated Panel Activities

In the past 30 years, there has been considerable research related to formula-based fund allocation programs. Findings and recommendations made two decades ago remain timely and relevant (see Box 6-1 for a summary). To make additional progress, the Panel on Formula Allocations will address both broad and focused issues relevant to developing, implementing, and evaluating federal and state programs.

A broad view will be provided by the panel's already initiated summary of goals, inputs, and formula features for the universe of federal funds allocation programs listed in the General Services Administration's Catalogue of Federal Domestic Assistance. Similar summaries will be considered for selected state programs.

To increase understanding of formula allocation processes and of the role of Congress and federal or state agencies in developing and administering formula-based fund allocation programs, the panel plans to commission a series of papers on some or all of the following topics:

1. *Retrospective case studies of the evolution of formula allocations for specific programs.* The universe for these studies would be formula allocation programs which have been in operation for a substantial period of time. The focus of each study will be on how the formula and the allocation process have changed over time, the reasons that changes were made, and evaluation of their effects on equity, efficiency, or other appropriate measures of program effectiveness. See the list of relevant program features

> **BOX 6-1**
> **Previous Recommendations on Allocation Formula Design**
>
> - Specify program goals and statistics clearly and completely, avoiding underspecification of program goals and over-specification of formula inputs.
> - Promote an effective collaboration between the legislative branch and statistical agencies to ensure that the best information, properly analyzed, is used for formula inputs.
> - Improve the frequency and quality of formula performance testing and monitoring to ensure that formulas are neither too sensitive to short-term changes nor too insensitive to long-term changes.
> - Promote communication about undesirable formula practices and data problems.
> - Avoid the use of funding thresholds by specifying a gradual transition from receiving no funds to receiving the "above-threshold" amount.
> - Evaluate the effectiveness of specific versus general and simple versus complex proxies for the characteristic that policy makers want to use for targeting program funds.
> - Develop additional statistical series for use in fund allocation.
> - Use comparable state-specific information when allocating funds to states.
> - Evaluate the costs and benefits of improving the accuracy of estimates of formula inputs and undertake efforts to improve accuracy only when justified by such an evaluation.
>
> SOURCE: U.S. Office of Statistical Policy and Standards (1978:25-28) and Spencer (1982:528).

below for additional details. A commissioned paper may cover a single program or more than one program. Programs that might be covered fall into three groups:

(a) U.S. federal formula allocation programs. Some good candidates might include Elementary and Secondary Education Act Title I, the Special Supplemental Nutrition Program for Women, Infants, and Children (WIC), the State Children's Health Insurance Program, the Substance Abuse and Mental Health Services Administration (SAMHSA) Block Grants, Medicaid, Highway Planning and Construction, Employment and Training Assistance, and Special Education Grants to States under the Individuals with Disabilities Education Act.

(b) State programs, i.e., programs in which state governments use formulas to allocate state funds to counties, municipalities, and other jurisdictions. A case study might focus on a particular state or cover similar programs in two or more states. State aid to education is an area of considerable current interest.

(c) Foreign and international formula allocation programs. Examples of foreign programs would be Canada's system for fiscal equalization among its provinces or for allocating revenues from "harmonized sales taxes" among the federal and three participating provincial governments. At the international level, several agencies, such as the United Nations, the International Monetary Fund, and the U.N. Development Programme, use formulas to allocate aid funds.

2. *Prospective case studies of specific programs or groups of programs.* These studies would cover areas in which significant developments are expected to occur during the projected life of the Panel on Formula Allocations. Possible examples are:

(a) The reauthorization process for the Temporary Assistance for Needy Families block grant program. The program will be coming up for reauthorization soon, providing an opportunity for a sort of anthropological/organizational study, tracking and analyzing the roles of congressional staff, federal, and state agency staff and other players in the process. The paper would describe the initial allocation procedures established under the Personal Responsibility and Work Opportunity Reconciliation Act of 1996 and subsequent changes, including any that may be included in the reauthorization legislation. High-performance bonuses are a special feature that may merit attention.

(b) Revision of New York state's formula for the allocation of state education funds. A recent court decision has declared the current formula to be unconstitutional and has given the legislature until September 2001 to revise it (the case may be appealed). This paper might be expanded to cover similar recent developments in other states.

(c) A study of several continuing programs whose allocation formulas rely on decennial census data. Describe in detail how the transition from 1990 to 2000 census data takes place. Are new legislation or regulations needed? Which year's allocations are the first to be affected? How is the transition affected by hold-harmless provisions? What can be learned about how to improve future transitions when new decennial census data are released (or are replaced by American Community Survey data)?

3. *A quantitative analysis of historical trends in U.S. formula allocation programs: 19xx to 2000.* Using data from the Federal Assistance Awards Data System (FAADS) and other sources, develop annual time series data for the amounts of federal funds distributed to states and other recipients through formula allocation programs. To the extent possible, provide data classified by type of program, type of recipient, and other salient characteristics. An initial conceptual analysis will be needed to determine the scope of programs to be included and define appropriate classification variables. The FAADS data go back only to 1981; other sources will be needed to extend the series farther back, and there may be problems in achieving comparability among different sources.

4. *An analysis of long-term effects of formula allocations on the legislative process.* A historical analysis of how the introduction and increasingly widespread use of formula allocation processes has helped national and state legislatures in their functioning, for example, by shifting the language of the debate.

5. *Alternative measures of fiscal capacity.* Measures of fiscal capacity or capability are often used in allocation formulas to represent the possibility of a recipient area meeting its needs from state, local, or private funds. The most commonly used measure has been and still is per capita income. In 1989, in response to a congressional requirement, an alternative measure, total taxable resources, was developed by the U.S. Treasury Department's Office of Economic Policy for use in the allocation formula for the SAMHSA block grants. This study would evaluate these two alternative measures, comparing their suitability from a conceptual point of view, the quality and timeliness of the data sources used, and other relevant features. The evaluation might include comparisons of the effects of using the two measures in specific formula allocation programs, such as Medicaid.

6. *An empirical analysis of the effects of hold-harmless provisions.* Building on the work of Zaslavsky and Schirm (2000), describe the various kinds of hold-harmless provisions that have been used and how they interact over time with changes in the total amounts appropriated for the program and with other formula features, such as thresholds. Include some real examples showing what the allocations would have been with and without hold-harmless provisions.

7. *The rationale for hold-harmless and threshold provisions in formulas.* For the most part, allocation formulas are based on measures of need, capacity, effort, and cost that are more or less directly related to program goals. Other formula features, such as hold-harmless provisions and thresholds, are often added on the grounds that they are needed for reasons of administrative efficiency, for example, to avoid disruptions caused by large changes in the amounts received from one year to the next. The purpose of this study would be to undertake a detailed analysis of the rationale for such formula provisions, examine how they vary among programs, and, if possible, gather empirical evidence about how effectively they are meeting their stated objectives. As part of the study, the experiences and views of local program administrators should be sought.

8. *Procedures for combining different components of allocation formulas.* Many formulas incorporate different components representing need, capacity, effort, and costs, although it is challenging to determine how best to combine these components in a single allocation formula. Examine a large number of allocation formulas with multiple components to identify methods used to combine them. Develop a typology of alternative methods and identify their advantages and disadvantages.

9. *The role of the public comment process in the development of allocation formulas.* In those instances in which some features of the allocation process are determined by regulation rather than legislation, the proposed regulations have to be published in the *Federal Register* for public comment. Identify some instances in which this has occurred, analyze the volume, nature, and sources of comments received, and analyze changes to the proposed regulations as a result of the public comment process.

10. *Measuring the effects of the statistical properties of input data on the achievement of program goals.* Several studies have examined the effects of

the statistical properties of input data, such as sampling variance, persistent bias, and lack of conceptual fit, on formula allocations and on the degree of equity achieved by the process. However, very little is known about how such errors affect the achievement of program goals. The purpose of this paper would be to develop prospectuses for one or more experimental or quasi-experimental studies designed to measure gains in efficiency that might result from investment in the development of more accurate input data. Specific programs would be identified for the proposed experiments and design protocols proposed.

11. *Facilitating analysis and interpretation of simulations of alternative allocation formulas.* Simulations of alternative formulas and processes are frequently used in the development of new programs and the revision of existing allocation formulas. Given the complexity of many formulas and the sometimes unexpected ways in which formula elements and features interact, it may be difficult, even for statisticians, to evaluate the outputs of simulations. This paper would attempt to identify summary measures and graphical outputs that would make it easier for interested parties to understand the properties of alternative formulas and make choices among them.

Each of the case studies (topic 1 above) will describe the establishment and subsequent development of one or more formula allocation programs. If more than one program is included, the programs will be selected either because they have similar objectives and target populations or because they illustrate significant contrasts in approaches to formula allocation.

The relevant program features include:

- program goals;
- target population, if applicable;
- a brief description of services provided by the program;
- first-level recipient units, e.g., states, counties, metropolitan areas;
- the formula and its elements, including need, capacity, effort, and cost;
- information, from legislative history and other sources, about considerations that affected the formula's initial development;
- sources of input data used to estimate formula elements;
- other formula features, such as thresholds, minimums, and hold-harmless provisions;

- research studies designed to measure and suggest possible improvements in the equity or efficiency of the allocation process;
- studies designed to evaluate impacts of the program; and
- the relationships between legislators, program agencies, recipient units, and other interested parties in the operation of the program.

The primary focus of each case study will be on changes in these program features during the life of the program. What specific changes occurred, why were they made, and how did they affect the achievement of program goals? Other commissioned papers will address topics such as alternative measures of fiscal capacity, the rationale for and effects of hold-harmless and threshold provisions in formulas, and procedures for combining different components of formulas. For a selected set of programs, the panel intends to study statistical properties involving formula inputs, features, and outputs, and to compare actual allocations to those that would have resulted from using alternative formulas or processes.

The panel anticipates developing examples of good and poor practice, including consideration of allocation programs that do not use formulas and programs that have changed after an evaluation (e.g., WIC). To provide background for possible changes in the properties or types of input data as new data sources become available in the next few years, the panel will document the principal data sources currently used in the major formula based allocation programs. The panel will also identify information gaps and potential new information sources.

Fund allocation programs are extremely complicated systems. Their design, implementation, and evaluation are very specialized activities and require understanding of diverse fields. Not surprisingly, there is a knowledge gap. To help bridge this gap, increase public understanding, and increase the effectiveness of formula-based fund allocation programs, the panel plans to develop a handbook that addresses issues in formula program development, implementation, and evaluation (Box 6-2). The target audience includes members of Congress and their staffs, federal and state policy makers and program administrators, and other interested parties, such as advocacy groups. Principal topics include explaining the complex interrelations between inputs and formula features, necessary a priori and ongoing evaluations, and the properties of the most common data sources. The handbook will aid in developing new allocation formulas and evaluating existing formulas. Issues related to appreciating, accommodating, and communicating uncertainty will receive substantial attention in the handbook.

**Box 6-2
A Handbook on Fund Allocation Formulas:
Preliminary Table of Contents**

1. Introduction
    1.1 Fund allocation formulas: an overview
        1.1.1 An early example: the Morrill Act
        1.1.2 General Revenue Sharing
        1.1.3 A statistical summary of current programs
    1.2 The parties involved
        1.2.1 The Congress
        1.2.2 Program agencies
        1.2.3 First-level recipients
        1.2.4 Individual beneficiaries
    1.3 Alternative approaches
        1.3.1 Amounts specified in legislation
        1.3.2 Specific formula in legislation
        1.3.3 Goals in legislation; formula developed by program agency with public comment
    1.4 Types of formula allocations
        1.4.1 Closed mathematical statements
        1.4.2 Iterative procedures
        1.4.3 Matching and cost-sharing provisions
    1.5 Purpose of the *Handbook*
    1.6 Intended audience
    1.7 Uses of the *Handbook*
        1.7.1 Developing a new formula
        1.7.2 Periodic allocations
        1.7.3 Analyzing an existing formula

2. Program goals
    2.1 Target population
    2.2 Services provided
    2.3 Desired outcomes

3. Basic formula features
    3.1 Target allocation units
        3.3.1 Multilevel allocations
    3.2 Frequency and timing of disbursements
    3.3 Provisions for administrative costs
    3.4 Program rules

4. Components of formulas
   4.1 Measures of need/workload
   4.2 Measures of funding capacity
   4.3 Cost differentials
   4.4 Effort
   4.5 Interactions among components

5. Special features of formula allocations
   5.1 Thresholds and other eligibility criteria
   5.2 Minimum and maximum values
   5.3 Hold-harmless provisions
   5.4 Interaction of special features with size of and changes in program appropriations

6. Data sources for estimating formula components
   6.1 Decennial censuses
   6.2 Household surveys
   6.3 Other statistical programs
   6.4 Administrative records
   6.5 Factors to consider in choosing data sources
       6.5.1 Conceptual fit
       6.5.2 Level of geographic detail available
       6.5.3 Timeliness
       6.5.4 Quality
       6.5.5 Costs of collecting new data or processing existing data
   6.6 Combining data sources to produce model-based estimates
   6.7 Updating estimates

7. Special topics
   7.1 Step functions v. continuous functions
   7.2 Hold-harmless provisions v. moving averages

8. Operational considerations
   8.1 Steps in developing a new formula
   8.2 Quality assurance procedures
       8.2.1 Replication
       8.2.2 Analysis of change from prior years
   8.3 Evaluating a formula
       8.3.1 The use of simulation techniques
       8.3.2 Exploratory data analysis

References and Bibliography

NOTE: Papers prepared for the Committee on National Statistics—marked by an asterisk (*)—are available from the committee, 2101 Constitution Ave. NW, Washington, DC 20418.

Adams, J., and A. Burnam
    1995   Federal block grants: What do statisticians have to contribute? In Window on Washington, Constance F. Citro, Ed. *Chance* 8(4):40-42.

Barro, S.M.
    1993   *Federal Policy Options for Improving the Education of Low-Income Students.* Volume III, Countering Inequity in School Finance. Santa Monica, CA: RAND.

Bixby, L.E.
    1977   *Statistical Data Requirements in Legislation.* Report prepared for the Committee on National Statistics, National Research Council. Washington, DC: National Academy Press.

Blumberg, L., J. Holahan, and M. Moon
    1993   *Options for Reforming the Medicaid Matching Formula.* Washington, DC: The Urban Institute.

Break, G.F.
    1980   *Financing Government in a Federal System.* Studies of Government Finance: Second Series. Washington, DC: The Brookings Institution.

Burnham, M.A., P. Reuter, J.L. Adams, A.R. Palmer, K.E. Model, J.E. Rolph, J.Z. Heilbrunn, G.N. Marshall, D. McCaffrey, S.L. Wenzel, and R.C. Kessler
    1997   *Review and Evaluation of the Substance Abuse and Mental Health Services Block Grant Allotment Formula.* RAND Drug Policy Research Center, MR-533-HHS. Santa Monica, CA: RAND.

Center for Governmental Research
    1980   *Formula Evaluation Project: Final Report.* Report to the National Science Foundation. Rochester, NY: Center for Governmental Research.

Center for Urban and Regional Study
  1975 *Alternative Formulas for General Revenue Sharing.* Report to the National Science Foundation. Blacksburg, VA: Virginia Polytechnic Institute and State University.
Ellett, C.
  1976 A study of data requirements for population-based formula grants. *Statistical Reporter* 77: 48-57.
  1978 *Analysis of the Federal Medical Assistance Percentage (FMAP) Formula, Volume I: Executive Summary.* Annandale, VA: JWK International Corporation.
Fay, R.E., and R.A. Herriot
  1979 Estimates of income for small places: An empirical Bayes application of James-Stein procedures to census data. *Journal of the American Statistical Association* 78: 269-277.
Ghosh, M., and J.N.K. Rao
  1994 Small-area estimation: An appraisal. *Statistical Science* 8(1): 55-93.
Goodnough, A.
  2001 New York City is shortchanged in school aid, state judge rules. *The New York Times*, January 11.
Maurice, A.J., and R.P. Nathan
  1982 The census undercount: Effects on federal aid to cities. *Urban Affairs Quarterly* 17(3): 251-284.
Moskowitz, J., S. Stullich, and B. Deng
  1993 *Targeting, Formula, and Resource Allocation Issues: Focusing Federal Funds Where the Needs Are Greatest.* A Supplemental Volume to the National Assessment of the Chapter I Program. Washington, DC: U.S. Department of Education.
Nathan, R.
  1980 The politics of printouts: The use of official numbers to allocate federal grants-in-aid. Pp. 331-342 in *The Politics of Numbers*, W. Alonso and P. Starr, Eds. New York: Russell Sage Foundation.
National Research Council
  1980 *Estimating Population and Income of Small Areas.* Panel on Small-Area Estimates of Income and Population, Committee on National Statistics. Washington, DC: National Academy Press.
  1995 *Measuring Poverty: A New Approach.* Panel on Poverty and Family Assistance Concepts, C.F. Citro and R.T. Michael, Eds., Committee on National Statistics. Washington, DC: National Academy Press.
  1997 *Small-Area Estimates of Children in Poverty, Interim Report 1, Evaluation of 1993 County Estimates for Title I Allocations.* Panel on Estimates of Poverty for Small Geographic Areas, C.F. Citro, M.L. Cohen, G. Kalton, and K.K. West, Eds., Committee on National Statistics. Washington, DC: National Academy Press.
  1998 *Small-Area Estimates of Children in Poverty, Interim Report 2, Evaluation of Revised 1993 County Estimates for Title I Allocations.* Panel on Estimates of Poverty for Small Geographic Areas, C.F. Citro, M.L. Cohen, and G. Kalton, Eds., Committee on National Statistics. Washington, DC: National Academy Press.
  1999a *Small-Area Estimates of Children in Poverty, Interim Report 3, Evaluation of Revised 1995 County and School District Estimates for Title I Allocations.* Panel on Estimates

of Poverty for Small Geographic Areas, C.F. Citro and G. Kalton, Eds., Committee on National Statistics. Washington, DC: National Academy Press.

1999b *Equity and Adequacy in Education Finance: Issues and Perspectives.* Committee on Education Finance, H.F. Ladd, R. Chalk, and J.S. Hansen, Eds. Washington, DC: National Academy Press.

2000a *Small-Area Income and Poverty Estimates: Priorities for 2000 and Beyond.* Panel on Estimates of Poverty for Small Geographic Areas, C.F. Citro and G. Kalton, Eds., Committee on National Statistics. Washington, DC: National Academy Press.

2000b *Small-Area Estimates of School-Age Children in Poverty: Evaluation of Current Methodology.* Panel on Estimates of Poverty for Small Geographic Areas, C.F. Citro and G. Kalton, Eds., Committee on National Statistics. Washington, DC: National Academy Press.

Orland, M.E.
1988 Relating school district resource needs and capacities to Chapter I allocations: Implications for more effective service targeting. *Educational Evaluation and Policy Analysis* 10(1): 23-36.

PricewaterhouseCoopers LLP
2000 *Effect of Census 2000 Undercount on Federal Funding to States and Local Areas, 2002-2012.* Washington, DC: PricewaterhouseCoopers LLP.

Schirm, A.L.
1995 *State Estimates of Infants and Children Income Eligible for the WIC Program in 1992.* Washington, DC: Mathematica Policy Research.

Smith, W., and A. Parker*
2000 An Overview of Formulas for Allocation of Funds. Paper prepared for the Workshop on Formulas for Allocating Program Funds, Committee on National Statistics, April 26-27, National Research Council, Washington, DC.

Spencer, B.
1980a *Benefit-Cost Analysis of Data Used to Allocate Funds.* New York: Springer.
1980b Implications of equity and accuracy for undercount adjustment: A decision-theoretic approach. Pp. 204-216 in *Conference on Census Undercount: Proceedings of the 1980 Conference.* Washington, DC: U.S. Census Bureau.
1982a Technical issues in allocation formula design. *Public Administration Review* 42:524-529.
1982b Concerning dubious estimates of the effects of census undercount adjustment of federal aid to cities. *Urban Affairs Quarterly* 18:145-148.
1985 Avoiding bias in estimates of the effect of data error on allocations of public funds. *Evaluation Review* 9:511-518.

Stanford Research Institute
1974 *General Revenue Sharing Data Study.* Menlo Park, CA: Stanford Research Institute.

Straf, M.
1981 Revenue allocation by regression: National Health Service Appropriations for Teaching Hospitals. *Journal of the Royal Statistical Society, Series A* 144(1): 80-84.

Stutzer, M.
1981 Parametric Properties of Tax Effort Revenue Sharing. Federal Reserve Bank of Minneapolis, Document SR-86.

U.S. General Accounting Office
- 1983 *Changing Medicaid Formula Can Improve Distribution of Funds to States.* GAO/GGD-83-27. Washington, DC: U.S. Government Printing Office.
- 1991 *Formula Programs—Adjusted Census Data Would Redistribute Small Percentage of Federal Funds to States.* GAO/GGD-92-12. Washington, DC: U.S. Government Printing Office.
- 1991 *Medicaid—Alternatives for Improving the Distribution of Funds.* GAO/HRD-91-66FS. Washington, DC: U.S. Government Printing Office.
- 1996 *Federal Grants Design Could Help Federal Resources Go Further.* AIMD-97-7. Washington, DC: U.S. Government Printing Office.
- 1997 *School Finance—State Efforts to Reduce Funding Gaps Between Poor and Wealthy Districts.* GAO/HEHS-97-31. Washington, DC: U.S. Government Printing Office.
- 1998 *School Finance—State and Federal Efforts to Target Poor Students.* GAO/HEHS-98-36. Washington, DC: U.S. Government Printing Office.
- 1999 *Formula Grants: Effects of Adjusted Population Counts on Federal Funding to States.* GAO/HEHS-99-69. Washington, DC: U.S. Government Printing Office.

U.S. General Services Administration
- 1998 *1998 Formula Report to the Congress.* Washington, DC: U.S. General Services Administration.

U.S. Office of Statistical Policy and Standards
- 1978 *Report on Statistics for Allocation of Funds.* Statistical Policy Working Paper 1. Federal Committee on Statistical Methodology, Subcommittee on Statistics for Allocation of Funds. Washington, DC: U.S. Department of Commerce.

Zaslavsky, A., and A. Schirm*
- 2000 Interaction Between Survey Estimates and Federal Funding Formulae. Paper presented at the Workshop on Formulas for Allocating Program Funds, Committee on National Statistics, April 26-27, National Research Council. Washington, DC.

Appendix A

Workshop Agenda and Participants

**WORKSHOP ON FORMULAS FOR ALLOCATING
PROGRAM FUNDS**

Wednesday, April 26, 2000, Green Building, Room 104

8:30   Continental Breakfast

9:00–9:15   Welcome and Opening Remarks
            Andrew White, Director, Committee on National Statistics
            Tom Louis, Workshop Chair, University of Minnesota

9:15–10:45 **SESSION I, Introduction and Workshop Overview**
            The purpose of this session is to provide the workshop participants, who come from a variety of government and academic venues, with an overview of the manner in which federal program funds are allocated to jurisdictions based on statistical formulas. This session will address the background of formula allocation, types of programs covered, trends in amounts of funds allocated, current distribution of funds across departments, data sources, and previous studies of statistical features of formulas.
            Presenter: Wray Smith,
               The Harris Smith Institutes   *9:15-9:45*
            Discussant: Jerry Fastrup, U.S. General
               Accounting Office   *9:45-10:00*

	Discussant: David McMillen, U.S. House Government Reform and Oversight Committee	10:00-10:15
	Discussant: Martin H. David, University of Wisconsin	10:15-10:30
	Floor Discussion	10:30-10:45
10:45–11:00	Break	

11:00–12:30 **SESSION II, Title I Case Study**

This session will provide an overview of the Title I Education funding formula. The impact of updated census data on allocations to school districts will be examined, and issues the Education Department faced in using new data to allocate Title I funds to school districts will be addressed.

Presenter: Sandy Brown, U.S. Department of Education	11:00-11:25
Presenter: Graham Kalton, Westat	11:25-11:50
Discussant: Bruce D. Spencer, Northwestern University	11:50-12:10
Floor Discussion	12:10-12:30

12:30–1:30 Lunch

1:30–3:00 **SESSION III, Features and Consequences of Formula Allocation**

This session will explore some of the consequences for federal formula allocation when particular features, such as thresholds and hold harmless, are present in the formula. In particular, the effect on allocations of changes in measurement systems, such as changes in precision and frequency, in the presence of these features will be examined.

Presenter: Alan Zaslavsky, Harvard University	1:30-1:55
Presenter: David Betson, University of Notre Dame	1:55-2:20
Discussant: Paul Siegel, U.S. Census Bureau	2:20-2:30
Discussant: Robin C. Fisher, U.S. Census Bureau	2:30-2:40
Floor Discussion	2:40-3:00

3:00–3:15   Break

3:15–4:45   **SESSION IV, Substance Abuse and Mental Health Services Formula Block Grants**
This session will outline the intention, as stated by Congress, of the substance abuse and mental health services block grant formula allocation. Alternative measures for the formula elements will be presented. Comparisons between the current formula measures and alternative measures will be discussed.

Presenter: Albert Woodward, Office of Applied Studies, Substance Abuse and Mental Health Services Administration	*3:15-3:40*
Presenter: John Adams, RAND	*3:40-4:05*
Discussant: Jane Maxwell, Texas Commission on Alcohol and Drug Abuse	*4:05-4:25*
Floor Discussion	*4:25-4:45*

4:45–5:00   General Discussion
5:15   Reception
6:30   Dinner

Thursday, April 27, 2000, Green Building, Room 130

8:30   Continental Breakfast

9:00–10:30   **SESSION V, WIC, Fund Allocation and Small Area Estimation in the WIC Program**
This session will provide an overview of the WIC formula, changes in estimate of WIC eligibles, the current method to produce these estimates, and goals for future estimates.

Overview of WIC: Ronald Vogel, Special Nutrition Programs, U.S. Department of Agriculture	*9:00-9:25*
Presenter: Allen L. Schirm, Mathematica Policy Research, Inc.	*9:25-9:50*
Discussant: David Betson, University of Notre Dame	*9:50-10:10*
Floor Discussion	*10:10-10:30*

10:30–10:45 Break

10:45–12:00 **SESSION VI, Roundtable Discussion**

The panel members will be state officials, representatives from federal statistical agencies, and researchers of formula allocation. The discussion should be oriented toward the future of available data and estimates for formula allocation. What issues should be taken into consideration for future data, estimated inputs, and features of formula allocation?

Panel Members: (15 min. each followed by 15 min. of discussion)
Paula J. Schneider, Principal Associate Director for Programs, U.S. Census Bureau
Linda Gage, California Department of Finance
Katherine K. Wallman, Chief Statistician, U.S. Office of Management and Budget
John E. Rolph, Chair of the Committee on National Statistics, University of Southern California

- Are there problems with the quality and timeliness of available data?
- Are there features of new and future datasets that are particularly relevant to issues of formula allocation, such as timeliness and level of aggregation?
- Do you think that the estimates used in formulas or the features of formulas have unintended consequences with respect to equity between jurisdictions? If so, what changes might resolve such problems?
- Do you have any suggestions for changing formulas, data, and estimation procedures?
- What issues could be usefully addressed by the Committee on National Statistics in a study of statistical and data needs for allocation formulas?

12:00–12:30 Concluding Remarks
Henry Aaron, Brookings Institution

12:30 Lunch

PARTICIPANTS

Presenters

Henry Aaron, Brookings Institution, Washington, DC
John Adams, RAND, Santa Monica, CA
David Betson, Department of Economics, University of Notre Dame
Sandy Brown, U.S. Department of Education
Martin David, University of Wisconsin
Jerry Fastrup, U.S. General Accounting Office
Robin Fisher, U.S. Census Bureau
Linda Gage, California Department of Finance
Graham Kalton, Westat, Rockville, MD
Thomas Louis, RAND, Alexandria, VA
David McMillen, U.S. House of Representatives
Jane Maxwell, Texas Commission on Alcohol and Drug Abuse
John Rolph, Marshall School of Business, University of Southern California
Allen Schirm, Mathematica Policy Research, Inc.
Paula Schneider, U.S. Census Bureau
Paul Siegel, U.S. Census Bureau
Wray Smith, The Harris Smith Institutes
Bruce Spencer, Department of Statistics, Northwestern University
Ronald Vogel, U.S. Department of Agriculture
Katherine Wallman, U.S. Office of Management and Budget
Albert Woodward, Substance Abuse and Mental Health Services Administration
Alan Zaslavsky, Department of Health Care Policy, Harvard University

Invited Guests

Craig Abbey, Congressional Research Service
Chip Alexander, U.S. Census Bureau
Todd Barrett, U.S. Department of Agriculture

William Bell, U.S. Census Bureau
Ken Bryson, U.S. Census Bureau
Lynda Carlson, National Science Foundation
Cheryl Chambers, U.S. Census Bureau
Stephen Cohen, Bureau of Labor Statistics
Paulette Como, Congressional Research Service
Mike Compson, U.S. Treasury Department
Thomas Corwin, U.S. Department of Education
Robert Dinkelmeyer, U.S. General Accounting Office
John Eltinge, Bureau of Labor Statistics
Deborah Fulcher, Substance Abuse and Mental Health Services Administration
Linda Ghelfi, U.S. Department of Agriculture
John Guyton, PricewaterhouseCoopers
Stefan Harvey, Center on Budget and Policy Priorities
Daniel Kasprzyk, National Center for Education Statistics
Jerry Keffer, U.S. Census Bureau
Victoria Lazariu-Bauer, New York State Department of Health
Bette Mahoney, The Harris Smith Institutes
Don Malec, U.S. Census Bureau
Marge Martin, U.S. Department of Housing and Urban Development
Marge Miller, Congressional Budget Office
Albert Parker, Synectics for Management Decisions, Inc.
Jeff Passel, Urban Institute
Ken Prewitt, U.S. Census Bureau
Charles Roberts, Substance Abuse and Mental Health Services Administration
Susan Schechter, U.S. Office of Management and Budget
Kathleen Scholl, U.S. General Accounting Office
Susan Sieg Tompkins, Congressional Budget Office
Marjorie Siegel, U.S. Department of Housing and Urban Development
Tom Slomba, U.S. General Accounting Office
Ian Soper, U.S. Department of Education
William Sonnenberg, National Center for Education Statistics
Paul Strasborg, U.S. Department of Agriculture
Doug Williams, National Center for Health Statistics
Michael Williams, U.S. General Accounting Office
Josh Winicki, U.S. Department of Agriculture
Peyton Young, Johns Hopkins University

Committee on National Statistics Staff

Andrew White, *Director*
Heather Koball, *Study Director*
Connie Citro, *Study Director*
Shelly Ver Ploeg, *Study Director*
Thomas B. Jabine, *Consultant*
Jamie Casey, *Senior Project Assistant*

Appendix B

Biographical Sketches of Panel Members and Staff

THOMAS A. LOUIS (*Chair*) is a senior statistical scientist at RAND and adjunct professor of biostatistics at the Johns Hopkins University School of Public Health. His research interests focus on Bayesian methods with applications in health, environmental, and public policy. He is coordinating editor of *The Journal of the American Statistical Association*, a member of the Committee on National Statistics, on the board of the Institute of Medicine's Medical Follow-up Agency, and on the executive committee of the National Institute of Statistical Sciences. He was on the IOM Panel to Assess the Health Consequences of Service in the Persian Gulf War and was on the CNSTAT Panel on Estimates of Poverty for Small Geographic Areas. He is a fellow of the American Statistical Association and of the American Association for the Advancement of Science. He received a Ph.D. in mathematical statistics from Columbia University and from 1987 to 1999 headed the department of Biostatistics at the University of Minnesota.

GORDON J. BRACKSTONE is assistant chief statistician responsible for statistical methodology, computing, and classification systems at Statistics Canada. From 1982 to 1985 he was the director-general of the Methodology Branch at Statistics Canada, and previously he was responsible for surveys and data acquisition in the Central Statistical Office of British Columbia. His professional work has been in survey methodology, particularly the assessment of the quality of census and survey data. He is a fellow of the American Statistical Association and an elected member of the International Statistical Institute. He received B.Sc. and M.Sc. degrees in statistics from the London School of Economics.

DANELLE J. DESSAINT (*Project Assistant*) is a staff member of the Committee on National Statistics. Her projects include ones on formula allocations, State Children's Health Insurance program, elder abuse, and institutional review boards. She has a B.A. in communications from Wingate University and formerly worked as an editor at Tribune Media Services in Glens Falls, NY.

VIRGINIA A. de WOLF (*Study Director*) is a senior program officer on the staff of the Committee on National Statistics. Previously, she has worked at the U.S. Office of Management and Budget, the Bureau of Labor Statistics, the National Highway Traffic Safety Administration, the U.S. General Accounting Office, and the University of Washington (Seattle). In the early 1990s she served as the study director of the panel that authored *Private Lives and Public Policies: Confidentiality and Accessibility of Government Statistics*. Currently, her areas of research interest are confidentiality and data access as well as statistical policy. She has a B.A. in mathematics from the College of New Rochelle and a Ph.D. from the University of Washington (Seattle) in educational psychology with emphases in statistics, measurement, and research design.

LINDA GAGE is California's state demographer and chief of the Demographic Research Unit at the California Department of Finance. She has held various positions within the Demographic Research Unit since 1975. Previously she held research and teaching assistant positions at the University of California. Her fields of demographic activity are in applied demography, small-area data analysis, migration, race/ethnicity, population estimates and projections, analysis of U.S. Census Bureau programs/procedures/data, fertility, and mortality. She has an M.A. in Sociology from the University of California, Davis, with emphasis in demography.

MARISA A. GERSTEIN (*Research Assistant*) is a staff member of the Committee on National Statistics. She is currently working on projects on welfare impacts, WIC, and elder abuse. Previously, she worked at Burch Munford Direct, a direct mail company, and the National Abortion and

Reproductive Rights Action League. She has a B.A. in sociology from New College of the University of South Florida.

HERMANN HABERMANN is director of the Statistics Division for the United Nations. Previously, he was the deputy associate director of the U.S. Office of Management and Budget, where he also served as chief statistician. In addition to his knowledge of statistics, he brings to the committee his knowledge of the federal statistical system. He has a Ph.D. from the University of Wisconsin, Madison, in statistics.

THOMAS B. JABINE is a statistical consultant who specializes in the areas of sampling, survey research methods, statistical disclosure analysis, and statistical policy. Recent clients include the Committee on National Statistics, the National Center for Health Statistics, and several other statistical agencies and organizations. He was formerly statistical policy expert for the Energy Information Administration, chief mathematical statistician for the Social Security Administration, and chief of the Statistical Research Division of the U.S. Census Bureau. He has provided technical assistance in sampling and survey methods to several developing countries for the United Nations, the Organization of American States, and the U.S. Agency for International Development. His publications are primarily in the areas of sampling, survey methodology and statistical policy. He has a B.S. in mathematics and an M.S. in economics and science from the Massachusetts Institute of Technology.

ALLEN L. SCHIRM is a senior fellow at Mathematica Policy Research, Inc. Formerly, he was Andrew W. Mellon assistant research scientist and assistant professor at the University of Michigan. His principal research interests include small-area estimation, census methods, and sample and evaluation design, with application to studies of child well-being and welfare, food and nutrition, and education policy. He is currently an Associate Editor of *Evaluation Review*. He served on the Committee on National Statistics Panel on Estimates of Poverty for Small Geographic Areas and is currently a member of its Panel on Research on Future Census Methods. He is a member of the American Statistical Association's Section on Survey Research Methods Working Group on Technical Aspects of the Survey of Income and Program Participation. He has an A.B. in statistics from Princeton University and a Ph.D. in economics from the University of Pennsylvania.

BRUCE D. SPENCER is a professor of statistics at Northwestern University. He chaired the Statistics Department at Northwestern from 1988 to 1999 and 2000 to 2001. He directed the Methodology Research Center of the National Opinion Research Center (NORC) at the University of Chicago from 1985 to 1992. From 1992 to 1994 he was a senior research statistician at NORC. At the National Research Council he was a panel member with the Mathematical Sciences Assessment Panel, and the Panel on Statistical Issues in AIDS Research. He served as study director for the Panel on Small Area Estimates of Population and Income. He received his Ph.D. from Yale University.